96 £3

Peter Richardson

4 December, 197

Oxford.

HACIENDAS, PLANTATIONS AND COLLECTIVE FARMS

Haciendas, Plantations
and
Collective Farms

Agrarian Class Societies—Cuba and Peru

JUAN MARTINEZ-ALIER

FRANK CASS : LONDON

First published 1977 in Great Britain by
FRANK CASS AND COMPANY LIMITED
Gainsborough House, Gainsborough Road,
London E11 1RS, England

and in the United States of America by
FRANK CASS AND COMPANY LIMITED
c/o International Scholarly Book Services, Inc.
Box 555, Forest Grove, Oregon 97116

ISBN 0 7146 3048 9

Printed in Great Britain by
Page Bros (Norwich) Ltd, Norwich.

Contents

Preface

The studies collected here explain the organization of haciendas in highland Peru — especially sheep farming haciendas before the current land reform — and the organization of sugar cane farming in Cuba, both before and after the revolution of 1959. I call these studies essays because they are not the detailed monographs which could be written on the material accessible in Cuba and Peru. However, the interpretations I present do rest on a solid empirical base. They also consider theoretical issues whose relevance goes beyond the recent agrarian history of Cuba and Peru. In the introduction the common themes are brought together and their theoretical ramifications are further developed.

For access to archival materials I owe a considerable debt to Julio LeRiverend, acting head of the Cuban National Archives, and, in Peru, to the historians involved in the setting up of the Centro de Documentación Agraria — Heraclio Bonilla of the Instituto de Estudios Peruanos, and Pablo Macera of San Marcos University. I am also indebted to those who worked with me at the C.D.A. (particularly Beatriz Madalengoitia and Humberto Rodríguez) and to the Peruvian land reform authorities, especially Dr. Guillermo Figallo, president of the Tribunal Agrario.

I should like to thank Magnus Mörner for sending me advance drafts of his bibliographical articles on tenant-labour in Latin America, on the agrarian history of the Cuzco region, and on the historiography of the Andean haciendas (later published in the *Hispanic American Historical Review,* May 1973). To Eric Hobsbawm I am grateful for letting me see the draft of an article on land invasions. I am very much aware of my debt to the work of Manuel Moreno Fraginals and Juan Perez de la Riva on Cuban agrarian history, and I have benefited greatly from the work of Pablo Macera, Rafael Baraona and Cristobal Kay on Andean haciendas.

These essays were written while I was a research fellow of St. Antony's College, Oxford and I am grateful to the college

for its support over nearly ten years and for leave to spend
more time in absence than in residence. The Introduction was
written in 1974, while I was teaching at the Universidade
Estadual de Campinas (S. Paulo, Brazil), and final corrections
and improvements have been made at the Universidad
Autónoma de Barcelona and in Cerdanyola del Vallès (Barcelona), where I am at present doing research on peasants who
have become industrial workers.

Some of these studies have been published before: the first
in the *Journal of Peasant Studies,* I : 2, January 1974; the
third in *Oxford Agrarian Studies,* II : 1, 1973; a preliminary
version of the fourth appeared in *St. Antony's Papers,* 22, 1970,
Oxford University Press. The second paper is a revised version
of a paper given at the symposium on Landlord and Peasant
in Latin America held in Cambridge, December 1972. Quotations can be checked against the original Spanish versions in
Cuba, economía y sociedad and *Los Huacchilleros del Perú*
(Paris : Ruedo Iberico).

I

Introduction

Land Rent, Exploitation, Surplus

This book deals with different systems of land tenure and use of labour. My main line of enquiry is the study of the markets for land and labour. I should like, therefore, to begin by discussing topics such as 'exploitation' and 'surplus'.

There are two lines of argument against equating rent payment and extraction of surplus, of quite different pedigree — the 'substantivist' argument and the orthodox neoclassical economic argument.[1]

The 'substantivist' argument is succintly expressed in Dalton's analogy, that if a truly religious person pays tithe to his church this would hardly qualify as extraction of surplus since he would feel he was getting eternal salvation in exchange. There is then no objective theory of exploitation or surplus extraction. People are exploited if they themselves feel they are exploited, and not otherwise. Wolf divides the peasant's surplus (defined as income left over subsistence needs) into a reproduction fund, a ceremonial fund, and a fund of rent, of which only the last one would be extracted by a exploitative relation (Wolf, 1966). The 'substantivist' argument would be that it is difficult to distinguish (except subjectively) between the ceremonial fund and the rent fund. For instance, in cases where the church was also the landlord, rent payment might be disguised as a contribution to the parson's wellbeing. The distinction would also be difficult in cases where the peasant became a client and the landlord a patron, the client getting in exchange for his rent not only the use of a piece of land but also (perhaps mainly) a godfather for his children, or a political protector, or simply a pat on the back making him feel good. The economic relation would be embedded in the social organization, and no understanding would be gained by trying to disentangle the economic from the socio-religious-political aspects. Parenthetically, I should make clear that I am not a ready believer in patron-client relations.[2] Nevertheless, Dalton's view that extraction of surplus has no meaning if one leaves aside 'the subjective reactions of those who are forced to pay toward the politico-religious-economic superior to whom they pay' (Dalton, 1972:413) is telling against Wolf, since Wolf's division between a ceremonial fund and rent would have to depend on subjective evaluations.

In societies where, as the 'substantivists' put it, the market (including the market for land) becomes the integrating principle of social organization, it becomes easy to identify rent. Here rent payment would no longer appear as one side of a reciprocal obligation by virtue of which incommensurable things are exchanged. Here rent would be payment for the use of land. Wolf, however, regards 'rents not merely as an economic payment for the use of land, but as a payment made by virtue of relations of power that in the case of peasants, but not in the case of primitives, are guaranteed by the state' (Wolf, 1972:411). My feeling is that Wolf resorts to a political explanation before the economic issues have been properly explored.

Wolf's reference to rent as an 'economic payment' is at first sight surprising since in economic theory (since Ricardo) 'rent' is precisely a payment with no economic function. However, after the 1870s, the orthodox economists came to think of land rent as the remuneration to this factor of production which, under conditions of free mobility of factors should equal its marginal product value for an efficient allocation. Therefore, Wolf is perfectly entitled to speak of land rent as an 'economic payment'.

Since land is hardly immediately available for different uses (say, coffee or cotton growing, or agricultural and urban use), there might easily be in land rent a large part of 'economic rent' (for the relevant range of alternative uses). The orthodox neoclassical economic theorist would not object to equate this part of rent to 'surplus' and 'exploitation' (this is precisely the sense in which Pigou used the word 'exploitation'). The more inelastic the supply of land, the higher this element of 'economic rent' in land rent.

Thus, the orthodox neoclassical economist would, first, exclude as irrelevant the political element (except as it might impinge upon the mobility of factors of production), and would then criticize Wolf's statement that *all* rent payment is extraction of surplus, since a lot of rent payment is a remuneration to this factor of production which measures its contribution to production and which is needed in order to keep it in its most productive use (or to induce it to change its use). For instance, in location theory (of the Von Thünen variety) rent is clearly an economic payment; in the linear programming elaboration of location theory, there appears a set of location rents or payments for land which are consistent with transport costs and with the optimum spatial distribution of the programmed kinds of output.

Nevertheless, from a theory of derived demand and derived prices for a factor of production, it is illegitimate (even inside the neoclassical economic orthodoxy) to draw conclusions about personal (as distinct from functional) remuneration (of landowners, in this case), i.e. the theory does not legitimate payment of rent to landowners, but to land. The non-economist is entitled to be baffled. For, in real life, rents are not accounting devices which society (or a central planning agency) use in order to allocate economic activities spatially. They imply real transferences of income from rent-payers to rent-receivers.

Up to now, my main concern has been to show how the words 'rent', 'surplus', and 'exploitation' mean different things to different theories. What would they mean to the modern critics of neoclassical (or counter-classical) economic theory? By examining this modern critique (Dobb, 1973) we might find grounds for siding with Wolf, and against Dalton.

Note that in the modern critique of neoclassical economic theory, capital does no longer appear as a 'real substance' or resource the supply of which would require a market price (i.e. a rate of interest or profit), but rather as a social relation which endows capitalists with the ability to impose a mark-up profit margin over wage-costs, not so much by virtue of their (naked) political power as by virtue of their being capitalists, i.e. of their holding of ownership rights. This is quite similar to Wolf's definition of rent, though I am not quite sure how one should interpret Wolf's statement that rent is extracted by virtue of political power. In principle, there is no reason why extraction of rents by landowners should need more (naked) political power than extraction of profits by capitalists. Perhaps one should assume that land-ownership is generally more illegitimate than factory-ownership.

One specific Marxist problem is to explain how extraction

of surplus is possible in a society based on free contractual relations, where values are determined in the market. In cases where rents are set freely and contractually, Wolf's definition of rent as 'a payment made by virtue of political power' sounds faulty or, perhaps one should say, premature: rent is the economic expression of a social relation of production which is maintained, as Wolf says, by political power (including in this political power not only the police and the army, but also the more or less legitimate authority of the law and the courts, and the ideological hegemony which makes most peasants accept most of the time the landlords' claim to rent).

Even if one agrees that what appears as an economic relation between 'things' (land and money; or land and a share of produce; or land and hours of work) is really a socio-political relation between men, it is important to investigate to what extent the market (for land and labour) fulfils functions which under other systems (serfdom proper, or slavery, or the jajmani system) are fulfilled by force or by religion. This is the main point against Wolf.

Against the neoclassical economic orthodoxy, the main point is that the determination of rents (or profits) by the market does not exclude the objective existence of exploitation. The intrusion of monopoly (or, equivalently, the restriction to the mobility of peasants) would aggravate such exploitation (but note that land rent is not necessarily a 'monopoly price' — there are usually very many landlords, and, besides, peasants do migrate to the urban sector).

Let us now go back to the 'substantivist' argument against an objective theory of exploitation. Let us assume for the sake of the argument that while in Latin America rents are determined through the market (whether 'monopolistic' or not), in India 'the cost of renting land varies in inverse proportion to the rank of the tenant's sub-caste' (Mousnier, 1973:33). Clearly, in India incommensurable things would be exchanged in such a transaction: the economy is embedded in the social organization. But the economy is also embedded in the social organization in Latin America, and elsewhere (and *not* because of the existence of patronage or *compadrazgo* between landlords and peasants).

Dalton was not yet aware in 1972 of the modern critique of economic theory, since he argues, against Wolf, and in favour of his 'subjective' theory of exploitation, that there is no objective difference between a capitalist and a socialist system:

'If one asserts, with Marx, that workers in capitalist factories

are exploited because the market price of what they produce
is greater than their wages, then one should assert the same
for workers in communist factories, where the (centrally
planned) market price of what workers produce is also
greater than their wages. It is elementary economics that
wherever net investment is new capital goods is taking place,
or wherever the government is providing services — defense,
justice — which are not charged for directly, the take-home
wage will be less than the market price of what they
produce? (Dalton, 1972:414).

The modern critique does not interpret Marx in this simple
way. What it asserts is that the distribution of ownership and
the resulting income distribution are logically prior to price
determination, i.e. that the prices that serve to aggregate production of the various goods, and the profit rates that serve to
put a value on and then to aggregate the different means of
production to give as a sum the value of 'capital', all depend on
the existing distribution of income, and while it is true that in
any economic system growth requires savings and investment,
in the capitalist system the capitalists consume part of their
profits, and investment is directed towards lines indicated by
prices which in turn depend on income distribution. Both via
its effects on the values of 'capital goods' (which depend on
the share of profits), and on the pattern of demand (and
therefore on the valuation of the labour employed in different production lines), the unequal income distribution characteristic of capitalist relations of production tends to give to
labour less than it would otherwise get — capitalism is in this
objective sense an exploitative system, and the neoclassical
orthodoxy appears as the ideology of the capitalist class which
hides this fact by showing (unfortunately, in a circular way)
that factors' incomes measure their respective contribution to
production. In a socialist system, by definition, planning and
not the market is the basic economic mechanism; socialist
planners, even if they believe in Lange's 'market socialism' are
aware that they have to introduce from outside the strict economic system the desired product-mix, or the desired income
distribution, and the desired rate of accumulation. (Alterna-
tively, if they are not real socialist planners, they might introduce outside prices, such as world market prices).

The 'substantivist' school of economic anthropology will
have to incorporate the fact that in order to explain the formation of prices in a pure market system there is a need to introduce extraeconomic considerations about the ownership of

resources and the distribution of income. The market economy
is also embedded in the social organization — nevertheless,
there is always much to be gained by trying to disentagle the
economic from the socio-political-religious aspects. It seems to
me that neither Dalton nor Wolf have yet understood this properly, though I am not certain whether Wolf sees the relation
between the extraction of the peasants' surplus and political
power as an immediate one or rather as one mediated through
the specific relations of production.[3]

Such questions will again crop up at several points, especially
in the next essay (where it might seem that I subscribe to a
'subjective' theory of exploitation), and in the last essay (where
it might seem that I complain about Fidel Castro's lack of grasp
of economics).

Chayanov's Rule

Sahlins' recent valiant attempt to provide a new objective
theory of surplus may be mentioned here (Sahlins, 1972).[4]
This most ingenious attempt is based on what Sahlins calls
Chayanov's rule, i.e. that the higher the workers/consumers
ratio in the family, the less will the workers work. One assumption, of course, is limited needs. Surplus is defined as production arising from work over that expected according to
Chayanov's rule, i.e. if one observes domestic units whose
workers work harder and produce more than normal for their
workers/consumers ratio, then one is entitled to say that they
are producing a surplus (which they give to political rulers, or
which they might use themselves to become politically powerful, through gift-giving, for instance). What is 'normal' is established statistically; departure from such 'normality' is to be
attributed to the political sphere, and the actual distribution
will be different in different political organisations. Sahlins'
heuristic model of the 'domestic mode of production' shows
very clearly how the impulse to production comes often, in
tribal societies, from the political sphere.

However, there would be difficulties in extending Sahlins'
model to peasant societies. To start with, peasants are involved
in markets. Market transactions might be more or less peri-
pheral to their production decisions, but involvement in markets makes the assumption of limited needs perhaps less ten-
able, since they must be aware that there are things they are
too poor to buy. This objection, I would certainly agree, is not
decisive — the 'substantivist' school of economic anthropology
is mistaken in its assertion that economic theory requires an

assumption of unlimited needs; anybody at all familiar with
the debates on the economics of socialism and with the 'dual'
solution of linear programming applications would be aware of
this. Leaving then this point aside, the relevance of Sahlins'
Chayanov's rule to peasant societies is doubtful for other reasons. One can rephrase Chayanov's rule in a less paradoxical
fashion, as asserting that the higher the consumers/workers
ratio in each domestic unit, the more its workers will have to
work.

This would indeed be a relevant consideration in order to
approach, for instance, the study of female agricultural labour
in countries such as Brazil today or in Cuba after 1959. In
Cuba, it seems that as the threat of unemployment receded,
and as basic consumption needs were guaranteed through the
rationing system (this being equivalent to a decrease in the
burden of work needed to achieve a given standard of living),
the supply of agricultural wage labour decreased (both female
and male labour, in this case). In Brazil, the decrease in real
wages after 1964 has probably been a factor in forcing many
women to work in agriculture as wage labourers. Some kinds
of agricultural work are felt to be men's work, and nevertheless are nowadays done by women. A situation in which the
lower the real wage, the higher the labour supply (or, as in
post-1959 Cuba, the reverse situation) presents difficulties for
textbook economic analysis, since it is not at all clear what
marginal productivity under full employment conditions would
mean in such cases. One must conclude that within wide mar-
gins the rural wage must be set institutionally — but this, after
all, will not surprise any economist. Now, if Chayanov's rule
is given the second one of its alternative formulations, it becomes easy to link Chayanov to theories of overpopulation
and disguised (or open) unemployment, i.e. with the theory
of the labour surplus economy, not so much as a theory of
how to use this labour for economic development but rather
as a theory of why this labour is unused. Chayanov himself
was well aware of the possibility of explaining what might appear as the peasant families' periods of rest as being the consequence of lack of employment opportunities.

Collective agreements in Southern Spain between landowners
and labourers sometimes stipulated that married men would
have preference in employment over single men and over
women, i.e. that the higher the consumers/workers ratio, the
more often should the workers be given an opportunity to
work and earn their living. When public works were organized

in Southern Spain in times of seasonal unemployment (in the
same way as in the Brazilian Northeast), the rule was that those
with larger families would get preferential employment. Work-
sharing according to the principles implicit in Chayanov's rule
has also been reported by Jayawardena (1963). That there
might be mechanisms for regulating competition among peasants for work or land, when such competition would result
in a general and severe reduction in standards of living, is a
possibility that stands to reason. Of course, whether the unused
labour would or would not remain unused might well depend
on social organization and on the available techniques of production. The discussion on the use of idle labour for the formation of capital has often centred around the political requirements needed for this (Balogh, 1961).

Epstein's description of the economic principles that regu-
late the allocation of labour in the Hindu jajmani system makes
it clear how, in years where the crops fail, the distribution
takes place according to average productivity value, as opposed
to marginal productivity value (which, for full employment,
would be very low). There is no formal economic difference
with such systems as the *imponibile di mano d'opera* in Italy
or *alojamiento* in Southern Spain. However, in the jajmani
system competition is always absent, in good as well as in bad
times, because of religious principles. In Italy and Spain, such
systems came into being either because of the labourers' unions'
pressure, or as temporary arrangements while the Great Transformation to a pure wage-labour system and to an industrial
trade cycle (as opposed to an agricultural cycle) was proceed-
ing. The *imponibile* or the *alojamiento* (imposed by state or
municipal authorities) would be unnecessary in India (at least,
in Louis Dumont's India).

For many peasant societies, it would then be possible to
interpret Chayanov's rule as an adaptive strategy to lack of
employment opportunities (relative to the given socio-political
organization). However, this interpretation would not be plausible for tribal societies (or for sparsely settled peasants) practic-
ing shifting cultivation, and working very few hours a day.
There would be general agreement with Sahlins' view that under shifting cultivation people work little — they are poor, by
North Atlantic standards, but they do not seem to care, and
they do not use their ample free time for further work. This
is the 'paradox' which provides Sahlins with his starting point:
'the original affluent society'. Boserup gives an economic explanation of this 'paradox' which takes us a bit further than

Sahlins, and which is perfectly compatible with the assumption of limited needs. Productivity per hour might well decrease with intensity of cultivation (assuming no introduction of improved seeds, fertilizers, etc.), and therefore 'the careless idler of the old times, who had just to scratch the land to get enough food' (Boserup, 1965:61) will become a hard working peasant only when the increased population density will force him to do so. Should population density increase even further (and this was surely the situation in many parts of Eastern Europe between the wars), the peasants (and labourers) will again become part-time idlers, though scarcely 'careless' about it. In order to feed their families, they will perhaps compete with each other for work or land, or they will perhaps manage to restrict competition: the rationality of either course of behavior is best explained by conventional economics.

The Valuation of the Peasant Family's Labour, Economics of Scale, and Externalities

One of the themes that recurs in this book is that of the relations between economic and socio-political organization. Economic organization is described by using the terminology of conventional economics. On occasion, I have used some economic terms somewhat loosely. For instance, I sometimes refer to the superior economic efficiency of small scale farming (a point frequently made in these essays) as being the result of the existence of diseconomies of scale, while in fact this superiority arises not from the influence of scale as such (i.e., where a proportional increase in the quantities of all inputs by a multiple a would increase production by a multiple less than a) but from the cheaper supply price of self-employed than of hired labour.

This notion of a cheaper supply price came to me from two sources. First, Kautsky's idea that socialization would be more difficult in agriculture than in industry because of the survival powers of peasant farming, based upon the self-exploitation of its labour. Second, the theory of A. Lewis about there being two sectors in the labour market (one operating on the logic of fixed labour resources, the other working according to the market principles) — this theory gave rise to a large amount of literature on the 'labour surplus economy'.

Two persistent ideas in recent writings about peasants are that the peasant economy is based upon the commitment of family labour and that peasant farming has a special non-market rationality. Sometimes — as for instance in Shanin

(1971, 1973) — these two aspects appear in a list of 'traits' which, taken together, amount to a definition or description of what peasants are (as opposed, say, to farmers or to agricultural labourers). These are not analytically independent 'traits'. As suggested in the previous section, lack of employment opportunities outside the peasant holding would make it economically rational to commit the family's labour to the peasant enterprise, even when the return to such work is less than the ruling market wage.

The supply of labour does not only depend on the price of labour (whether in a normal or a 'perverse' manner) but also on the method of remuneration. Different land tenure and use of labour arrangements can be discussed in terms of their impact on labour supply. For instance, in my work on Southern Spain (1965, 1968, 1971) I showed how sharecropping (or small cash-renting) are a means to make the labourers (turned into peasants) work harder and better for only a small increase in total remuneration over that of wage labourers, thus increasing landowners' profits-rents. This is also relevant to the debate on exploitation mentioned above. Leaving aside the question of whether the wage labourers were initially exploited or not, I think it is fair to say that under such circumstances the sharecropper will be more exploited than the wage labourer, and this in quite an objective sense, whether he feels it or not.

The situation would be similar to that of the time-worker in industry (since there is nothing specifically 'peasant' in all this) who became a piece-worker, the piece-rate being set in such a way that after much more effort he earned only the same or a little more than before, thus increasing his boss's profits. One might easily imagine an industrial 'putting-out' system producing similar results. Notice also that in this increased rate of exploitation there is no intrusion of a monopoly element — it is the labour market which produces it, given the fact that the supply of labour can go up and down considerably, for a given number of people of working age.

Why are proper industrial workers never turned into the equivalent of individual sharecroppers? The reason is that there are indivisibilities in industry, or rather, that it is surely easier to externalize economies of scale in agriculture than in industry.

For instance, a large sugar mill will obtain a much higher extraction rate than a small *trapiche,* the mill getting in the region of ten or fifteen kilogrammes of sugar for each hundred kilogrammes of cane, and the *trapiche* only some five kilogrammes. This is the reason for the increasing concentration of sugar

milling. On the agricultural side, two factors would be at work.
On the one hand, vertical integration of sugar cane production
would allow better coordination between sugar cane growing
and milling (it is immaterial whether the mill would control
the cane growers, or vice versa). This coordination is needed
because cane loses in sucrose content and thus in value when it
takes more than a few hours to bring it to the mill. An increased
scale of agricultural operations would facilitate coordination,
but it might well happen that cane growing unit costs would
increase because of difficulties in labour supervision and because
of the reduction in the incentive to work.

The essays on Peru discuss sheep farming in the Andean
highlands. One problem confronting the large haciendas was
how to control scab in neighbouring Indian villages, where the
shepherds' own sheep came from. One may see scab control in
the neighbouring villages as an external economy to the haciendas; indeed, the large sheep farming companies often tried to
make the neighbouring villages incur the cost of disease control,
inducing or forcing them to build dip baths for their sheep. At
other times, however, the large sheep farming companies
would build dip baths for the neighbouring villages at the companies' expense.

From the point of view of the Indian shepherd in highland
Peru (or from the point of view of the sugar cane grower in
Cuba) it would have been impracticable to build the minute dip
bath (or the minute trapiche) required for his own needs. Disease control (or the processing of sugar cane) are operations
where large economies of scale obtain — but such economies
can be made external to the small farm.

In the case of sheep farming, there are economies of scale in
so many different activities that the large sheep farm is economically more efficient. This is rarely the case in agriculture,
insofar as labour costs are a large share of total costs. Nevertheless, tractors may sometimes entail a saving in costs whatever
the level of wages because ploughing with tractors in deep clay
soils means a big saving not only in hours of work but also in oxen
or mules. Harvesters and herbicides are also becoming competitive at very low levels of wages. In theory, one might still believe in the viability of small scale farming through the cooperation of small farms for some agricultural operations; however,
when there are indivisibilities in most operations, the small
farm is doomed despite its cheaper supply of labour.

It is sometimes alleged that the very large farm has an economic advantage in that it is able to obtain cheap labour through

its monopsonistic position in the labour market. The pattern
of settlement and the density of employment in rural areas certainly mean that the rural worker will have fewer alternative
employment possibilities within daily travelling range than the
urban worker. But it seems to me that the burden of proving
the existence of monopsony (or oligopsony) in the agricultural
labour *market* (i.e. leaving aside for the moment the existence
of serfdom, or slavery, or effective debt-peonage) must rest
with those who assert its existence. Some elementary calcula-
tions will be helpful. Assuming that rural workers can travel
easily up to, say, five kilometres, then for a worker to be
limited to one farm as a source of daily employment, he would
have to have his dwelling place in the centre of a farm of nearly eight thousand hectares. Besides, he could very well go to
work to a distant farm and stay there not for the day, but for
a week, or for a fortnight, or for a few months (as peasants
from highland Peru did, when they went down to the coast for
work on sugar or cotton plantations).

Finally, another possible economic advantage of large scale
farming comes from its comparably easier access to credit, to
which one should add the access to technical information
which often comes with access to credit. However, this is irrelevant to a comparison between farming based upon self-
employed labour and farming based upon hired labour, since
self-employment does not necessarily imply small scale. Moreover, the idea that the *agricultural* side of a plantation is a
heavily capitalized enterprise is simply wrong. If and when it
becomes heavily capitalized, then the issue of how labour is to
be used (whether as wage labour, or sharecropping labour, etc.)
usually disappears, precisely because of the saving in labour
that the modern techniques imply. The topic with which I deal
in these essays is the study of alternative ways of using labour
rather than the transition to a capital-intensive agriculture,
though it is of course true that some modern techniques (improved seeds, fertilizers) do not imply less use of labour, rather
the reverse. In any case, the question would be whether the
cheaper labour supply of small scale farming is or is not offset
by its more expensive sources of credit.

The Articulation of Capitalist and Non-capitalist Sectors

The coexistence of different types of relations of production
inside one social formation has been recently studied by some
French Marxist anthropologists. Perhaps some readers will be

thankful for one or two examples of what this terminology refers to. For instance, the existence of a rural subsistence sector
in the South African economy allegedly has the function of
cheapening the 'cost of reproduction' of the wage labour force
employed in the mines or in the plantations. The subsistence
sector assumes 'functions that capitalism prefers not to assume
. . . the functions of social security' (Meillassoux, 1972:102).
In this particular instance (whose relevance for highland Peru
is clear) one would like to know whether alternative arrangements would not really come out cheaper for the capitalists
(such as, for instance, confiscation of the land in the subsistence
sector, and compulsory or wage labour), and whether, therefore, the statements from spokesmen for the capitalists who
assert this positive economic function of the subsistence
sector are not merely a rationalization of their own misgivings
when confronted with the herculean and dangerous task of de-
priving the whole indigenous population of access to land. [5]
Unless the costs and benefits of alternative arrangements are
spelled out, it is difficult to know to what extent the capitalists' complaints (or praises) about the existing arrangements
are to be taken as accurate descriptions of reality or as ideological statements.

Both in this book and in my previous work on Southern
Spain I have addressed myself to such questions, trying to
understand the economic and political motivation of both landlords and peasants.

The existence of plantations run with wage labour has often
been explained by the assumed economic superiority of the
big capitalist firm, which displaces the peasantry. The existence
of haciendas run with 'serf' labour has been explained either by
landlords' lack of economic motivation (because of lack of
markets) or by the larger profits-rents to be extracted in this
way (as compared to wage-labour or to share- or cash-tenancy).
I have already remarked upon the anomaly of the use of wage-
labour, since labour becomes cheaper under sharecropping or
cash-tenancy arrangements: a wage-labour plantation economy
cannot be taken for granted. An Andean hacienda cannot be
taken for granted either, since the so-called 'serfs' were less exploited than wage-labourers would have been; i.e. they got more
out of the haciendas by doing less work. Both in Cuba and in
Peru, the previous ways of looking at the agrarian question
served, as we shall see, useful ideological functions; it was convenient to believe that plantations would proletarize the Cuban
peasantry, and that Indian serfs in Peruvian haciendas were

terribly exploited and would be well advised to integrate themselves into the wage-economy and to learn to speak Spanish.

The advance of capitalism used to be thought to imply the
inexorable proletarization of the peasantry, and for instance
both the *colono* system of Andean haciendas and sharecropping
elsewhere were deplored (both by liberal reformers and by
socialists) as feudal or semi-feudal survivals. We are now perhaps
coming to a time when the positive functions of the peasantry
and in general of the non-wage sectors of the economy are
nearly taken for granted. The reality, however, is not so simple.
I find it difficult to agree with Meillassoux's explanation, that
peasant agriculture with its 'obsolete organisation is maintained
as long as possible by capitalism as a means of cheap reproduction of the labour force' (Meillassoux, 1973:89). (In later work,
Meillassoux deals with this question more carefully.)

Consider, for example, the jajmani system, the Italian
imponibile and the Spanish *alojamiento* which, as suggested earlier, are formally similar economic institutions, since all three
substitute the logic of fixed labour resources (i.e., the logic of
peasant or self-employed farming) for the logic of the labour
market. However, Italian and Spanish landowners complained
bitterly against such systems, while one would not expect
Hindu landowners to complain.

On the other hand, the greater efficiency of peasant farming,
based upon a lower valuation of self-employed than of hired
labour, makes it difficult to speak of its 'obsolescence'. Peasants
produce cheaper commodities (and not only cheaper children),
and this is why we often find capitalist large farmers taking in
small tenants, tenancy being a sort of incentive wage.

Finally, there are cases (such as highland Peru) where the
survival of the peasantry is more a matter of successful political resistance on their part than of capitalists' economic con-
venience. The so-called 'serf' system in Andean haciendas was
uneconomic (from the landlords' viewpoint), but the Indian
peasants managed to hold on to their plots of land by using
ethnic boundaries as an arm in the class struggle.

So, the procedure to explain the existence of peasant farming in a capitalist economy would be, first, to see to what extent it can be explained by economic principles (Kautsky's
notion of self-exploitation, or the economists' notion of a
difference between the private and opportunity costs of labour
are relevant here); second, to pay attention to the political or
religious levels.

Labourers, Peasants, and Farmers: Class Position and Politics

The essays on Cuba deal with farmers (the sugar cane planters) and with labourers, while those on Peru deal rather with land-lords and true peasants.

The Cuban sugar cane growers were divided into various strata, according to the amount of land they farmed and the amount of cane they were allowed to grow and sell to the mills under the quota system introduced in the 1930s. I pay more attention to the large than to the small cane growers, partly because of the sources I used (the minutes of the meetings of the cane planters' association) and partly because I find them amusing.

I characterize their political beliefs as 'bourgeois nationalism'. They were certainly not part of a 'national bourgeoisie', if by this one means a social class able to industrialize its country independently. But they were both bourgeois (i.e., anti-working class), and nationalists (i.e., anti-American). Both attitudes fitted well with their social position, between the agricultural labourers and the sugar mills (which were, in part, owned by American companies). Their main grievances were over credit and marketing; those who were tenant farmers also asked for a 'land reform', and in this they joined with the poorer peasants. Their grievances over marketing (specifically, payment for the cane sold to the mills), and their conflict with the agricultural labourers set them apart from the ideal-type peasant concerned above all with access to land to cover subsistence needs and whose main source of labour is not hired labour but he himself and his family.

What I have to say about the Cuban cane planters will, I think, be uncontroversial. Their political behaviour corresponded closely to their class position. But when I deal with Peruvian peasants and with Cuban labourers, I might be easily misunderstood. I might appear to be saying that peasants and labourers are very much the same thing: both Peruvian peasants and Cuban labourers wanted land, both formed or attempted to form unions. In fact, my point is that before one embarks on generalizations about the political sociology of peasants one must know a lot more about the specific class position of different types of peasants; i.e., their economic relations with landowners (if any), the different ways a surplus is (or is not) extracted; the manner in which such economic relations are perceived by the actors. Besides, a given class position does not determine a particular political behaviour; conjunctures

are as important as socio-economic structure; people may have
at the same time conformist, reformist and revolutionary ideas.
If I may refer once again to my work on Southern Spain, I
think that I showed in a convincing way how and why it was
that labourers felt at the same time that a *reparto* of the large
estates should take place and that work for the landowners
should be carried out according to morally obligatory standards of effort and quality *(cumplir)*. Dual consciousness is a
disturbing notion since it means that whatever happens (a
revolution as well as a counter-revolution) fits into the model.
Nevertheless, I think that dual consciousness exists among
those who feel the existing situation as illegitimate but put up
with it.

There is something unconvincing in the distinction between
the industrial proletariat as a class of 'high classness' (in the
sense that a great deal of its politics is explained by its class
position), and the peasantry as a class of 'low classness'
(Hobsbawm, 1973:5). After all, one third of the British proletariat votes for the Conservative Party, and one half (not less
remarkably) for the Labour Party. Hobsbawm views the peasantry (in its natural state, so to speak) as a class incapable of
carrying through a socialist revolution at the state level. I myself rather believe that the obstacles the peasantry finds to
acting politically are similar in kind to those found by an industrial proletariat: lack of coordination (aggravated because of
ecological more than sociological factors), the conformist side
of their dual consciousness, the need to find allies and the
lack of sufficient numerical strength, and the continuous, in-
tense repression from the police and the army.

The attitude with regard to the individual possession of land
is clearly relevant to the prospects for a socialist revolution.
One might be tempted to attribute a different meaning to the
Cuban labourers' and to the Peruvian peasants' desire for land.
In highland Peru, it might seem the expression of a conservative attachment to the integrity of the Indian peasant communities, whose land had been taken over by the estates. In
Cuba, land or work opportunities were seen as equivalent
alternatives in a proletarian struggle centred around seasonal
unemployment; access to land was as good as a secure, well-
paid job, because labourers (in the existing state of development of techniques) could have become viable peasants or
small farmers.

The revolution of 1936 in Southern Spain was fuelled at

the same time by a proletarian struggle against unemployment
and by 'old-fashioned', 'primitive' anarchist egalitarianism.
Both the proletarian and 'peasant' elements seem to me quite
compatible, except perhaps in the framework of a theory of
history which would see the industrial proletariat directed by
the communist party as the main agent of revolution in both
industrial or agrarian societies. 'Old-fashioned' egalitarian
anarco-syndicalism was also the creed of the industrial workers
of Catalonia who are the only workers (to date) in twentieth-
century Western Europe who have made a revolution — not,
certainly, an orthodox Leninist one.

Surely the position should be that both the industrial proletariat (or parts of it) and the peasantry (if not composed of
contented owner-operators) are potentially revolutionary
classes. The guidance of a party is useful to the peasantry, not
so much to awaken their class consciousness as to protect them
from immediate repression and to connect them in some way
to sources of real power (in this respect, the Maoist theory of
power coming from the barrel of a gun was vastly more practical than the anarchist doctrine of the evilness of all power).
Peasants are usually politically powerless not because they are
peasants but because they are poor.

There is another point to be made in a comparison between
proletarian and peasant politics. The Marxist concept of 'class'
goes together with 'class struggle' which in turn entails a conflict over the extraction of surplus through specific relations of
production. While 'industrial proletariat' refers to people in a
similar class position (though even here one should distinguish
between large scale and family industries, etc.), 'peasantry' is
a concept with less analytical bite. To talk of the 'peasantry'
as a class of 'low classness', without specifying whether the
'peasants' are owner-operators, or squatters, or sharecroppers,
or serfs, is a statement of such generality that the analogy
would be to say that the *urban* workers have much class consciousness — the first question would naturally be: which urban
workers? — white collar workers? independent artisans? industrial workers in large factories? workers in sweat-labour shops?
domestic servants? housewives?

The proper characterization of agrarian productive relations
also has important consequences for the analysis of agrarian
change trends (including land reform), which is the topic of
the last section of this Introduction. Before this, I wish to consider briefly another type of 'peasant' — the frontier squatter.

Squatters

A type of rural conflict which is both widespread in Latin America and politically important is that between squatters and would-be landowners. This is class conflict, and consequently the squatter does not fit into Shanin's conception of the 'peasant', whose main enemy is not so much the landowner as 'outsiders' belonging to the state political structure. The squatter, like the wage labourer, belongs to Shanin's 'analytically marginal' categories.

In the fourth essay I mention briefly the existence of *precaristas* in pre-revolutionary Cuba, partly explained by the reluctance on the part of landowners to take in proper cashtenants or sharecroppers because of the protection the tenancy laws provided for tenants. There were also another kind of *precaristas* in Cuba (in the sierras of Oriente province) who were similar to the squatters of the whole Amazon basin. They settled in public land (or in land deemed as public), cleared the land, and grew both subsistence and cash crops. They usually had to fight in the courts or in the fields against the so-called *geófagos* who after a few years appeared in the scene with genuine or forged property titles, ready to profit from the squatters' work.

This is in its essentials the conflict between *posseiros* and *grileiros* in many of the frontier regions of Brazil (see for instance Velho, 1972). Its ingredients are a fluid situation as regards title to land (usually public domain in the process of being privatized), and a divergency in the strategy of land use. The ecology of such regions would permit either the sparse settlement of squatters who would grow food crops under a system of long fallow, with high productivity per day of work because of the use of fire; or the land, after it has been cleared, can be devoted to cattle (or to permanent tree crops). The ecology prevents continuous cropping. Hence the possibility of the landowners claiming that land which, from the squatters' viewpoint, is recovering its fertility, growing bush and secondary forest after a couple of years of cultivation, is in fact abandoned. Where the land permits continuous cropping, then the squatters' eviction demands more brute force.

One Brazilian economist (Oliveira, 1973) has referred to such appropriation as primary accumulation of capital. In less fighting words, one could perhaps describe it, in Nurksian terms, as capital formation with underemployed labour. It is suggestive that the Peruvian squatters in the eastern slopes of the Andes are often called *mejoreros* (*mejorar* = to improve). *Mejoreros* who settle in land already privatized usually work under contracts which specify their obligations and remuneration. In one particular case I studied, that of the hacienda San Carlos, in Chanchamayo, Junín, the main provisions of the 1959 contract were as follows. Mejoreros received a plot of land for six years, and they had to clear it and plant avocado trees on it. Every year they had to perform several cultivation tasks, carefully specified: three *lampeos* and *limpieza de coronas* at a distance of two metres around the trees (weeding); elimination of *coquis* (a pest); fertilization and the biannual spreading of insecticides (given to them by the owner). How they were to plant the trees was also specified: depth of hole, distance between the trees, and so on. Once the trees were planted, mejoreros were allowed to grow food crops (only some kinds were allowed) and also to build huts. If a mejorero did not fulfil his contractual obligations, he was to leave without indemnity, leaving also the standing food crops. If authorized by the owner, a mejorero could transfer his plot to a third person. At the end of the six years' period of contract (when the trees would be fully grown), mejoreros had to leave and would get, as a *donación voluntaria,* one thousand soles each if everything was in perfect condition. In 1964 the first land reform law was passed in Peru. The law promised *feudatarios* (i.e. all types of small tenants) the property of the land, after complicated procedures. Some of the mejoreros were expelled in 1965, according to the provisions of the 1959 contract but against the 1964 land reform law. By 1969 they had organized themselves into a Frente de Defensa y Desarrollo de los Colonos y Nativos del Perené y Chanchamayo, and they asked the authorities that the estate be broken up. They wrote to the President of Peru, 'in support of your actions in favour of all the peasants of our Fatherland, as never before History has seen', and they explained that 'we, the feudatarios, have suffered the punishment of prison and of fire in our humble homes', and that they had been unable to get justice done to them because the owner had bribed the authorities and had a relative who was a lawyer in the local land reform office. They also complained that the owner had collected the avocadoes on his own account; this was in agreement with the clauses of the contract but it was now felt as a grievance.[6]

This case is instructive because, legally speaking, such mejoreros were not squatters. Nevertheless, whether the land is appropriated by landlords beforehand or afterwards, the latent con-

flict is always the same. In this particular case, the immediate
occasion for the conflict was the land reform law. Authors
who feel that rural struggles have to do with 'land hunger' more
than with unemployment have very properly emphasized this
type of conflict, characteristic of the immense frontier region
of the Amazon basin.

Because there is widespread perception of the explosive
nature of such situations, a good deal of attention has been
placed on the allegedly crucial support from the local precaristas to the guerrillas of Sierra Maestra (Wolf, 1969). This goes
usually together with an emphasis on evictions as the central
agrarian problem in Cuba (see below, p.128) and with an
exaggerated appreciation of the military aspects of the Cuban
revolution. Though it is true that the sierras of Oriente province were the last regions of spontaneous settlement in Cuba,
and though the legal status of a lot of land in that province
was certainly felt to be doubtful (Martinez-Alier, 1972:148),
I think that the course of the Cuban revolution depended
more upon the facts discussed in my essays (initial middle
class support; subsequent radicalization due to pressure from
the working class) than from the support given by the squatters
of Oriente province to Fidel Castro's military feats. Actually,
Fidel Castro in Sierra Maestra was a politician with a military
guard who waited for Batista's regime and army to disintegrate
politically. Until 1959, Fidel Castro was not so much the
'catalyzer' of a peasant rebellion as the catalyzer of the
nationalist middle class's discontent which brought Batista
down.

Land Reform and Collective Farming

In this concluding section, I shall touch upon a few practical
questions, such as the administrative decisions concerning what
was called *reserva de tierra* in Cuba and *mínimo inafectable* in
Peru (the area the landowner was allowed to keep inside his
expropriated estate); the criteria used to calculate the number
of beneficiaries who can get into any expropriated estate (for
instance, whether the number is calculated according to a
labour standard or to an income standard); the forms of land
tenure and use of labour adopted by the land reform (family
farms, cooperatives, collective farms).

Such options usually respond to the land reformers' ideology
about the existing agrarian structure and about the desired one.
Sometimes, however, really pragmatic considerations might

take the upper hand. For instance, land reform authorities who are in principle in favour of constituting family farms might well opt for the setting up of cooperatives (at least as a first step) given the bureaucratic effort which the surveying and delimitation of many small farms would entail. In the Peruvian highlands there is a trend towards forming cooperatives in agricultural estates or, occasionally, they are incorporated into larger SAIS (see below, p.89). From the social-christian corporativist views of the Peruvian government, one could have expected a preference for family farms, and for a time I thought that the trend towards cooperatives could be explained by administrative expedience, coupled perhaps with the need to counter the landlords' argument that it would be uneconomic to split up the 'demesne' part of their estates. I now think that the formation of cooperatives (and SAIS) in Andean haciendas has to be explained in the light of what we now know about the functioning of such haciendas. We now realize that the fundamental problem was how to increase the peasants' workload. Cooperatives and the like (SAIS, PIARS), whose managers are appointed from above, are a more suitable means to rationalize work practices than family farms would be.

Two general points are appropriate here. First, land reform should be studied in the context of previous agrarian conflicts, and not as a subject in itself. Second, a land reform option can respond to different land reform ideologies; nevertheless, the study of its technicalities is a good first approach to the social meaning of a land reform. Let us see for instance, in some detail, two Peruvian examples, one from the coast (the hacienda Pasamayo, in the Chancay Valley) and one in the sierra (the hacienda Quirupuquio, near Tarma), sources being in both cases the land reform files for each hacienda.

Pasamayo, an estate of about five hundred hectares of irrigated land, was used for cotton growing. It was the property of the Acuña family (one Acuña was married to one Graña, of the neighbouring and famous hacienda Huando). In the 1930s, Pasamayo had been rented as a unit to Tay Hermanos (not the well known Japanese firm which rented most of the Chancay Valley haciendas at the time). Tay had promptly subrented the whole of the estate to *yanaconas,* of whom there were some already in the 1920s and possibly before that. The land reform file for Pasamayo contains many contracts of *yanaconazgo* because under the 1964 land reform law *yanaconas* had to prove their tenant status. Though *yanaconazgo* is an institution which bears an ancient, pre-hispanic name, in the coast

it is really a form of tenancy well designed for cotton growing,
and which was not regulated by custom but by bargaining.
Thus one finds proposals from a contract of yanaconazgo from
the Sindicato de Yanaconas of hacienda Pasamayo, prior to
the tenancy legislation of 1947. Such proposals were that yanaconas would pay ten quintals of cotton, and not sixteen, for
each fanegada of land, because the land had many stones; also,
that interest would be paid on the *habilitación* (the money
advanced by the landlord) at the rate of one percent per month,
but that it would be paid in cash and not in cotton (this probably because of disputes on the price of cotton); another provision asked for cotton to be paid at the current market price;
also, twenty percent of the plot allocated to each yanacona
would be used for *panllevar* (food crops), according to the
legislation then in force, and the rent would be lower than for
the eighty percent used for cotton; yanaconas would be in any
case entitled to a minimum daily wage if they eventually made
less than this as sharecroppers or renters; finally, the firm was
asked to build a school. From 1947 onwards yanaconazgo was
regulated by law, and contracts are in all coastal haciendas
printed as official forms. There was however some variation
from hacienda to hacienda since clauses were added or deleted.
The 1947 legislation (enacted under president Bustamante) was
the result of debates on land reform, which had the usual outcome of producing legislation protecting tenants; this in turn
had a no less usual outcome: some tenants were dislodged, an
indemnity normally being paid to them, and not replaced by
others. This applies only to the coast, not to the sierra. However, a large part of hacienda Pasamayo was still *yanaconizada*
by the early 1960s, some three hundred hectares out of five
hundred. The debate on land reform of the 1960s, the consequence of the Cuban revolution, the Alliance for Progress, and
even more of the land invasions and formation of unions in
the sierra, resulted in the 1964 law which promised *feudatarios*
the property of the land they farmed.

Why were rentiers or sharecroppers such as the coastal yanaconas called feudatarios? Any connotation of lack of economic
rationality is doubly wrong, both because it is a mistake to
charge feudal landlords, unable to dismiss serfs, with economic
irrationality and, more relevantly, because wage-labour is more
expensive than tenant-labour in the absence of legislation on
rural tenancy. But, it seems clear that by the mid--1960s the
way left to defend agrarian capitalism was to attack agrarian
feudalism. If by feudalism one means restrictions to geographi-

cal mobility and customary, instead of contractual, arrangements, it is of course silly to include coastal yanaconazgo
under the feudal heading: yanaconas were like subcontractors.
Whether feudalism is or is not an adequate description of conditions in the sierra will be discussed in the first and second
essays: the tenancy arrangements (use of a piece of land in
exchange for labour services on landlord's land) would seem to
justify at first sight the use of the analogy with serfdom, while
at the political level *gamonalismo* has been used to denote the
assumed subordination of central power to local interests.

In hacienda Pasamayo, the land reform of 1964 had by
1969 effectively given the ownership of their plots to the sixty-
eight yanaconas left; the average size of plot was about 4.5
hectares. There were also over thirty permanent labourers who
worked mostly in the rest of the estate, and these men began
to write petitions to the president of the republic shortly after
the military coup of 1968, asking for a wider land reform:
specifically, the breaking up of the rest of estate and the allocation of one plot to each family. After a few months, petitions
were based upon article 45 of the land reform law of 1969,
which established that estates would be taken over without
indemnity and with no area reserved for the owners, if labour
legislation had not been complied with in the past. To show
that this was indeed the case, the Sindicato de Braceros of
hacienda Pasamayo (founded in 1957) wrote to say that the
landowner had not complied with specific points which had
been agreed upon in collective agreements negotiated in previous years. Petitions were signed by the Sindicato's Secretario
de Defensa, Secretario de Organización and Secretario General,
all of them labourers in the hacienda. Collective agreements
were negotiated after the annual *pliego de reclamos* was submitted by the union, dealing with two main types of questions:
wage increases and conditions of work. Improvements in work
conditions included, for instance, preference in employment
to the permanent labourer's sons over casual labourers from
outside the hacienda; the provision of first aid and medicines,
extensive to the labourers' families; that latrines were installed
and also drinkable water, properly piped in and not from the
irrigation canal; that machetes, sickles and axes be supplied by
the hacienda; that more houses be built; that electric light be
installed in the kitchens of the *rancherías;* also, that electric
light not be switched off from the union's meeting room until
the meetings were over; that a loan be granted to the union to
build a new meeting room; that loans be granted to labourers

in need; that the hacienda grow food crops, to sell to the
labourers at cost prices; that the hacienda donate powdered
milk to the labourers' children; that a small market place be
built on the hacienda, to sell everyday necessities; that each
labourer be provided with boots, once a year; that transport
on Sundays to Huaral, the neighbouring town, be arranged
and paid for by the hacienda.

To their own undoing the owners had repeatedly agreed to
some of these points, including latrines and drinkable water.
The estate was taken over in May 1970 because the agreements
had not been fulfilled. Article 45 became notorious for a time
since Pasamayo was only one of several coastal haciendas taken
over under its rules. After a while, a new procedure was introduced by which landowners could appeal to the Ministry of
Labour if article 45 was invoked by the land reform agency.
The Ministry of Labour then set up a court of inquiry, and
gave some delay to the owners to fulfil the agreements reached
with the unions, or the general provisions of the social legislation even if not specifically agreed to. If all haciendas without
latrines had been taken over wholesale and without indemnity,
the land reform would have proceeded swiftly. The Acuña
family complained that, after all, the town of Huaral also lacked
properly piped potable water.

Apart from the yanaconas and from the permanent labourers,
there was also a third party in Aucallama, a village nearby,
which also claimed rights to the land of Pasamayo. In December
1969 a group of forty men from Aucallama wrote to say that,
without detriment to the rights of feudatarios and labourers
of Pasamayo, they believed that land should also be made available to them, *campesinos sin tierras,* i.e. casual labourers. Apparently, they did not even get a reply. In some seasons of the
year, there had been around twenty casual labourers in Pasamayo. Finally, a cooperative was formed on the land not previously given to yanaconas. The cooperativists were thirty-
three permanent labourers and six employees, this excluding
the casual labourers. It is worth noticing that in one of the
pliegos de reclamos of former years the permanent labourers'
union of Pasamayo had generously asked that all casual
labourers be granted stability, i.e. be turned into permanent
labourers.

The number of beneficiaries in Pasamayo was determined
according to an income standard, each labourer being meant to
earn yearly about 30,000 soles (over 700 U.S.$; 1 U.S. $ =
42 soles). According to a labour standard, it appears that forty-
four labourers would be needed on average throughout the

year, according to the land reform engineers' settlement project, which disarmingly concludes: 'the deficit in available
labour (from the cooperativists) will be filled in by demographic
growth'. There are cooperatives in other cotton growing haciendas of the coast where, by contrast, the members are to earn
much less. Thus, in Sta. Rosa Huarangal and San José del
Monte, in Mala, Cañete, only thirty families could get into the
estate according to an income standard of only 13,500 soles
per year, but the number of actual beneficiaries was sixty-two
families, all of them headed by permanent labourers. This is
in contrast, for instance, to Casagrande, the showpiece of the
sugar cooperatives, where just the end of year bonus for 1970
was 8,000 soles. In Casagrande there are about 4,500 cooperativists. To each of them corresponds twenty-four hectares of
irrigated land, plus a share in the sugar-mill which must be more
or less equivalent to the value of another, say, twenty-five hectares of irrigated land. Though parts of the profits are paid out
in taxes, the cooperativists will have a standard of living far
higher than that of the surrounding masses of unemployed men
and women.

Before going on to consider one case which illustrates the
application of the land reform to sierra haciendas, I shall mention a few conflicts in the coast which do not appear in the
Pasamayo case. One is the conflict for land between yanaconas
and people who work for them, called *allegados*. They are
often relatives, which makes the conflict all the more acute.
Another is the conflict for land not only between permanent
and casual labourers, but between two kinds of casual labourers. Thus, in Cerro Alegre, Cañete, an association of resident
casual labourers was formed in April 1970 against the administrator's policy of bringing in casual labourers from the nearby
town, by lorryloads, by the day, instead of giving employment
to the casual labourers resident in the hacienda. In this same
hacienda, the permanent labourers' union wrote to the land
reform agency asking a pertinent question: whether female
labourers, their wives, were to be included on their own right
as beneficiaries of the land reform: they found no mention of
this in the law but as, in cotton growing, women work as much
as men, and as they foresaw there would be an excess of prospective cooperativists, they had an interest in defending
women's rights. They, however, in principle, did not like
women to work: 'it is only as a help in supporting our sons
and daughters that we see ourselves in the imperious need of
putting our wives to work'.

One conclusion must be that what now appears, under the

stimulus of the land reform, as a conflict for land or for rights
on land as a cooperative member, was before a conflict over
work opportunities. Because the legislation on unions did not
provide for unions formed on a town or regional basis, but for
one union in each coastal hacienda, the resulting inequities
(from the point of view of the poorest people) are bound to
be greater than if everybody had been previously unionized.

Let us now see one case of the application of the land reform
in the sierra, which will serve to introduce the main actors of
rural conflicts in the Peruvian highlands and, at the same time,
to reach some conclusions about the Peruvian land reform. In
December of 1969, the newspaper *Expreso* reported: 'More
than five hundred comuneros from Huacapo are demanding
the immediate expropriation of the hacienda "Quirupuquio",
the property of Celestino Vargas. The comuneros told us that
they themselves have only 2,500 hectares, insufficient to work
and live in . . . . The comuneros demand immediate action, as
it was taken in the haciendas of the coast' (the sugar plantations,
that is). According to the owner, the comuneros of Huacapo
had invaded part of the pasture land of his hacienda Quirupuquio, a hacienda of 930 hectares, comprised of seven hectares
of irrigated land, 60 hectares of non-irrigated agricultural land,
650 hectares of pasture land, and some 200 hectares of non-
usable land. This pattern of land use is interesting because
it is typical of many other mixed sierra haciendas: a tiny
piece of irrigated land at the bottom of the valley, and then
further up the slope some poor agricultural land, and then the
pasture land which comprises by far the largest area; but in
value the tiny irrigated and cultivated bits are of course very
important. This pattern of land use, and the ecological complementarity between the different uses, make it difficult to dis-
tribute sierra haciendas and to localize the area reserved to
the owners after expropriation.

This hacienda had seventeen feudatarios, according to the
land reform officials from whose reports this account is taken.
The so-called feudatarios got a daily wage, in November 1969,
of 30 soles, they also had rights to land, both agricultural and
pasture, and they had sheep in greater number than the landlord himself. In exchange, of course, they had to work the
landlord's agricultural land and to take care of his sheep. Their
standard of living must have been higher than that of the average landless or minifundist sierra peasant, and they must have
also expected some extra benefits from the land reform before
the threat from the communities outside materialized. Thus,

the feudatarios wrote to the land reform agency denouncing the invasion. The version of the conflict which seems to come closer to the truth is the record of an inspection made by land reform officials in January 1970 together with the *personero* of Huacapo and the owner of Quirupuquio. According to this version, the *comuneros* of Huacapo had long been using the 135 hectares of pasture land under dispute. It was only in September 1969 that the comuneros' peaceful use of that land was disturbed, and not by the landlord himself, but by some of the seventeen peasants of the hacienda who put their own sheep on that same piece of land.

The title to that piece of land was in doubt. The community of Huacapo had been recognized only in 1947, and its limits were set down on the basis of titles and a drawing going back to 1765 and 1809. The resolution which recognized Huacapo as a community mentioned explicitly that such recognition did not imply that the limits shown in that drawing were accepted. A proper plan had been made only in September 1969 by an official from the Ministry of Labour and Native Affairs, but the owner of Quirupuquio had objected to that plan. Therefore, the rights to the section of the hacienda under dispute, known by the names of Putacruz and Castilla Puquio, were still in doubt. And now, recently, some of the hacienda peasants, acting on their own volition or under promptings from the owner, had let their sheep loose in that area. The owner, in its turn, exhibited titles going back to 1693; yet, the hacienda was not registered in the Property Registry.

There was still another party to the conflict, the community of Cochas whose territory also bordered on Quirupuquio. They claimed they had not enough land to farm, because of population increase, and they said that they had stronger rights than the community of Huacapo. They said also that some of the feudatarios from Quirupuquio belonged to the community of Cochas: a person can be a peasant in a hacienda and a member of a community at the same time.

One conclusion on the Peruvian land reform is that there is not enough land to go round, given the Unidad Agraria Familiar (or Ganadera, as the case may be) standard applied both to the distribution of individual plots and to cooperatives, and given the number of potential beneficiaries. There are estimates showing that in the coast only one of every two potential beneficiaries will get land or become a cooperativist, and in the sierra only one in five. However, the land reform might be a success in that it will place agrarian conflicts at a lower, less

sensitive level. Peasants and labourers fight over the spoils, and
will go on fighting later over wages, employment, etc. since
both family farms and cooperatives will have to employ outside labour or will at least be expected to do so by the unemployed. This will have more of the nature of family quarrels
than of class conflicts between hacendados and communities,
or hacienda peasants, or labourers' unions, the old conflicts
which could more easily be a source of inspiration for analogous
conflicts in the industrial sector. For instance, Quirupuquio
and the neighbouring haciendas belonging to the same family,
were not large enough to fit all the feudatarios they had before.[7] This is a bit surprising since they all had made a living,
with something left over for the landlords. But, anyway, it is
clear that the communities of Huacapo and Cochas were disappointed. The conflict now will be between comuneros and
ex-feudatarios. In cases (such as the SAIS) where the land reform officials have substituted themselves for the previous
landlords, the conflicts remain, as we shall see, very much the
same ones.

There is no need to describe the Cuban land reform here,
since this is the subject of one of these essays. For comparative
purposes it is enough to say that the Cuban land reform of
1959 started off in much the same way as in Peru: a complex
bureaucratic operation geared to privilege a small sector of the
peasantry, at the same time leaving to the big farmers a substantial part of their farms (see below, p.116). It is instructive
to notice the priorities established by article 22 of the land reform law of 1959: the first priority was for peasants who had
been evicted in the past; then, for peasants of the region (the
country was divided into twenty-seven land reform regions)
who had less than two caballerías of land (27 hectares), who
would get extra land up to the two caballería limit; third, landless workers, who would become cooperativists provided they
had worked and *lived in* that same land; preference would be
given to peasants under two caballerías from other regions over
workers who did not fulfil these two conditions. In practice
this would have left a large majority of the active agrarian population outside the land reform. However, pressure from those
left outside, together with the radicalization due to the conflict
with the U.S. led in the end to a much wider land reform and
finally to collectivization of a large part of Cuban agriculture.
In 1960 work in Cuban cooperatives was still described in
terms which would be apposite for Peru: 'we can differentiate
two types of workers: those who are members of the coopera-

tives, who live and work always in the same work-place; those we may call workers for the cooperatives, casual labourers who do not live in them and who are part of that class of agricultural labourers who go from one work-place to other.'[8]

The last essay in this book deals with the problems of collectivization in Cuba rather than with its virtues, and perhaps we should emphasize the virtues here. The problem of unemployment was done away with (which a family farm or a cooperativistic solution, of an ample income standard, would have been unlikely to solve), and at the same time distribution of production took great steps towards equality. The costs of agricultural production (though not necessarily of the marketable surplus) have perhaps been higher (because of the reduction in effort) than they would have been under peasant farming or with cooperatives. In Eastern Europe (particularly in Hungary) collective farm managers are reinventing forms of tenancy as an incentive to work. This is related to the Cuban polemic on the use of 'material incentives', which is the subject of the last essay.

The Peruvian and Cuban examples show that 'land hunger' comes from unemployment or from the risk of unemployment or from ill-paid employment. 'Land hunger' is not a specifically 'peasant' feeling. 'Peasants' who are given good alternative employment lose their 'land hunger' (as has happened to millions of them in the industrialized countries), and, more relevantly to the present book, rural people of all descriptions (whether 'peasants', or labourers or squatters, and so on) desire land of their own in order to have an assured supply of food and an assured opportunity of work. They will cease to desire land if such things are assured to them in some other way.

What is in question is whether the land is an end in itself or whether land has an instrumental value, as a means of getting assured work and as a way to earn one's living. The Peruvian land reform movement of the 1960s chose to call peasants working in coastal or sierra haciendas 'feudatarios', in a tradition which goes back to Mariátegui and Haya de la Torre. The implication was that feudatarios would have preferential access to land. The aim of the land reform was not the socialist aim of destroying the labour market. The Peruvian left renounced this aim not because of tactical considerations but because until quite recently it was mistaken in its views on the agrarian systems of the country. The Cuban revolution, which began in the same way, developed quite differently; in my view, it would be too simple and rather inadequate to attribute this difference

to the fact that the Cuban rural population was largely composed of labourers while that of Peru (or at least of highland
Peru) was composed of 'peasants'.

The left considers a 'democratic' land reform as the suitable
objective when productive relations are of a 'feudal' or 'semi-
feudal' type, while a 'socialist' land reform is in order when
agricultural labour is in wage employment. Categories such as
'feudal' or 'capitalist' are ambiguous in cases such as, for instance, sharecropping, a form of use of labour which arises
from the unfettered operation of the land and labour markets,
but which is not wage-labour. On the other hand, sharecroppers
and also wage-labourers could well settle for a 'democratic'
land reform (which does not take care of the fundamental problem of unemployment), though they would also like a collec-
tivistic land reform which would give them assured employment and assured subsistence (possibly, peasants such as those
of highland Peru would share this view).

NOTES

1.   References to writings on 'surplus' by members of the substantivist
school of economic anthropology, in Dalton (1972).

2.   In my work on Southern Spain I showed how Pitt-Rivers' postulated
existence of a patron-client relation was merely a functionalist device
intended to lay a gloss over a conflict situation which centred around
unemployment in the labour market, and was not the ideological defence
of the system the landowners would choose themselves.

3. Wolf pays little attention to the threat of unemployment for
labourers *and peasants,* since he places all peasants in a pre-capitalist
world where extra-economic coercion is the rule.

4.   I was unaware of this work when I wrote these essays. My reference
to Sahlins in the first one alludes to his previous work on forms of
reciprocity.

5.   This type of explanation pays little attention to the difference between the functions and causes of a social institution. The argument is
similar to that which asserts that the nuclear monogamous family is helpful to the capitalist economy because the unpaid wife's and mother's
work allows capitalists to pay their workers less than they would otherwise pay. However, given the average size of working class family of most
industrialized capitalist societies, it seems that some kind of extended
family would cheapen even more the 'cost of reproduction' of the labour
force, freeing young mothers for the labour market.

6. The source is the *expediente de afectación* for this estate, i.e. the file compiled by the land reform agency.

7. The *expediente de afectación* for haciendas Quirupuquio, Yananyari, Casa Blanca, Ayas estimated the number of beneficiaries as follows: the number of *'unidades ganaderas familiares'* was 11, since the area in natural pasture was 3,467 hectares — each sheep would need one hectare, and each family needed 307 hectares. (The number of sheep was actually 5,500 before the land reform, but this was considered to be too many.) The number of *'unidades agricolas familiares'* was 27, since the agricultural area was equivalent to 256 hectares of type III (the various agricultural areas being converted to a common unit), and each family would need 9.5 hectares. Since the total number of feudatarios in these haciendas was 40, two families (40 − 38) could not become actual beneficiaries. In cases such as this, the land reform officials do in due course fiddle their *'unidades familiares'* until at least all of the resident population of the hacienda fits in.

8. The description of an official of the land reform region Camagüey—20, of 12 January 1960.

II

Peasants and Labourers
in Cuba and Highland Peru

This essay discusses some of the differences and similarities between a peasantry and an agricultural proletariat and considers the economic relations between peasants or labourers and landowners. It de-emphasises the specificity of 'peasant society' as an object of study both from the economic and the socio-political point of view.

The strongest claim to specificity has been made for Chayanov's so-called 'theory of the peasant economy', but I do not think this theory is of higher analytical value than conventional marginalist economics. Though there is perhaps no claim to a specific sociological and political theory of the peasantry, one often encounters a number of generalizations which command wide agreement, such as the denial of the applicability of 'social class' to the peasantry, which goes back to Marx's notion of the French peasantry as a 'sack of potatoes'. The low degree of 'organic solidarity' is often mentioned, too, since a peasant economy which is more or less a subsistence economy implies a low degree of division of labour. Peasant families like to have a plot of land and to grow their own food. They have little use for outsiders, especially for town dwellers. Politically, they are bound to be in a subaltern position; consequently, the less outside political interference there is, the happier they are. Unless strongly led from outside, the peasantry will not organize itself politically with the aim of influencing national politics. Its natural form of political action is inactivity, or spontaneous, disconnected uprisings directed towards acquiring or keeping rights to land. In contrast, agricultural labourers, together with other proletarians, will tend to combine and struggle, not for land but for higher wages and employment opportunities, and may provide the backbone of political parties whose aim is to conquer political power at the state level.

The assumed differences between a peasantry and an agricultural proletariat have led Shanin to call the latter a group 'analytically marginal' to the study of peasant society proper (Shanin, 1971). By throwing doubt on the existence of some of the assumed differences, this paper will contribute to the debate on whether the analysis of the peasantry requires special theoretical treatment.

Sharecropping and Wage-labour

We may start with Eric Wolf's notion of a peasant as a rural cultivator whose surpluses are transferred to a dominant group of rulers. After mentioning the surpluses which find their way into a replacement fund (seed for next year, for instance), and into a ceremonial fund (for wedding feasts, for instance), Wolf goes on to consider the surpluses which go into rent: regardless of whether that rent is paid in labour, in produce, or in money, where someone exercises an effective superior power, or *domain,* over a cultivator, the cultivator must produce a fund of rent (Wolf, 1966:9—10).

Though I share in the widespread admiration for Wolf's work, I find unsatisfactory, as regards economic relations, this notion of what a peasant is. One objection, which I shall consider only in passing, is that it leaves aside peasants who are owners of the land, or who manage in some other way not to pay rent. Wolf does briefly point out that a peasant owner's surplus can partly disappear, not through rent payment, but through the market. But this assumes imperfect (monopsonistic or oligopsonistic) markets (or some other theory of unequal exchange which I do not think Wolf really manages to develop). One may of course readily believe that markets are imperfect. One can also think of other mechanisms to extract part of the peasant owners' surplus, such as taxes. In this paper, however, I too shall leave peasant owners aside, not because they are unimportant in numbers in Southern Spain, pre-revolutionary Cuba, or highland Peru, but because my topic is agricultural labourers and peasants who work in or use resources of large estates or both. More specifically, my topic is the analysis of some agrarian class societies as social formations. I shall focus on the interplay between relations of production, class consciousness and conflict.

Another objection to Wolf arises from his inconsistency in applying the notion of transfer of surpluses to peasants only, but not to wage-labourers. A labourer, like a tenant, might think, 'we work, and the values we produce go mostly to the

landowner'. Andalusian labourers imply this in their saying
'bread should be buried deep in the earth, so that everybody
would have to dig it up by himself', i.e. everybody including
landowners. A sharecropper might say, 'we work, and 30 percent, or 50 percent, or 70 percent of what we produce, goes
to the landowner'. A *huasipunguero* in Ecuador or an *inquilino*
in Chile who, like the peasants in the estates of highland Peru,
were labour-service tenants, might have said, 'of our work, we
must give to the landlord one fifth, or one fourth, or one half'.
What matters, on the social plane, is that these people feel
exploited because they deny the right of landowners to get
part of the values produced. They deny, in fact, the landowners'
claim to land and the usefulness of their entrepreneurial functions. On one possible economic plane, what matters is whether the surpluses transferred — profits or rents — are just what
the labour market will bear, or are less, or are more. I shall
come back to this later on, by discussing restrictions (or rather,
lack of restrictions) on the geographical mobility of labour.

Rents can be regarded as prices for land which serve the function of allocating land to different uses and different people.
If rents, however, are taken as a means to extract the surplus
from peasants, then it becomes difficult to see at all clearly
the economic difference between labourers and at least some
types of tenants. In a sense, a labour theory of value is socio-
logically appropriate, from the labourers' point of view, at
least in my Andalusian experience: assuming one denies the
legitimacy of the landowners' claim to land, then landowners
may be seen as appropriating the surplus (be it as profits or as
rents) just because of their superior power.

It is also interesting to note that landowners themselves
decide between different forms of use of the labour available
partly in terms of the profits (or rents) they will get under one
system or the other. Thus they compare profits from using wage-
labour with rents from using sharecropping (or cash-tenancy)
and may decide accordingly. I do not wish to deny the importance of non-economic factors, and indeed non-economic factors play an essential role in my argument on landowners' decisions on the use of wage-labour, as will be seen later. But I
dispute the view that a changeover from customary to contractual relations in society means a changeover from sharecropping to wage-labour: 'an enormous proportion of the underdeveloped world's farm workers are sharecroppers: their relationship to the landowner is not contractual but customary'
(Warriner, 1969:45). Or consider, for instance, the following

statement: 'the shift to cash-workers is in most cases only partial. In São Paulo, for example, the sharecroppers and *colonos* in coffee plantations increased between 1955 and 1960 from 514,000 to 527,000, although at a slower rate than daily workers and piece workers whose number rose from 222,000 to 281,000. The former still outnumbered the latter by almost two to one in 1960 *although* São Paulo is considered to have one of the most dynamic agricultures in Brazil' (Feder, 1971:93).

Sharecropping, and also cash-tenancy, are forms of use of labour similar to carefully negotiated piece-work rates, not customary but variable. These are rationalized ways of using labour and they might be substituted for other forms — including wage payment — as agriculture becomes more commercialized, provided that the share of labour in total costs is still high.

The reasons, in economic terms, for the use of tenancy can be understood by placing oneself in the position of a landowner. By using wage labour, he has to pay wages, by definition greater than zero, while at least in some seasons of the year marginal productivity value under full employment conditions might fall below this wage. More important than the wasted unemployment in terms of man-days or hours can be the underemployment of the labourer's family, and the underemployment in terms of effort and quality of work, which landowners are unable to profit from in a wage-labour system. Piece-work is not used for many agricultural tasks precisely because the quality of work suffers. Thus, by turning labourers into tenants, landowners are able to get extra returns from the otherwise unemployed labour. The point at issue is to substitute a Chayanov type of peasant rationality for the rationality of the market price of labour. The peasant's labour input is not measured by himself against the wage level, but in terms of opportunity cost.

Here a short digression may be in order. When one says that the agricultural wage level is not determined by marginal productivity value, this can be understood in two senses. One, which I am using throughout, and which is the relevant one to the landowner confronted with decisions on use of labour and land tenure systems, is that the marginal productivity theory of wages explains, not the wage level, but the level of employment for a wage institutionally determined and which must still be accounted for in some way.

Secondly, there are economists who believe that in any economy, with or without full employment, the marginal productivity theory of wages does not explain the wage level. Simplifying a complex question (on which I am no expert), their critiques are as follows: The values that the commodities produced have in the market are not independent of the distribution of income, and therefore of the level of wages. Prices partly reflect unequally distributed purchasing power. One can explain the structure of prices (and therefore marginal productivity values) only if the distribution of income (and wage levels) which underlie it is taken as given. Also, one cannot explain the level of profits by the marginal productivity theory of capital, since the values of capital goods themselves depend on the level of profits, which in turn depends on the share of wages. Though I am aware of these fundamental challenges to the neoclassical theory of distribution of production among the factors of production, I think however that it is quite proper and uncontroversial to use the marginal productivity theory of wages, *not* to explain the wage level, but to approach the study of the landowners' decisions on employment and land tenure.

Though it might be slightly more controversial, in view of the way in which Chayanov's so-called 'theory of the peasant economy' is likely to be used in the controversies in economic anthropology, I would add that, in my view, Chayanov's theory is not in the least a new branch of economic theorizing, since his insights can easily be phrased in the language of conventional economics. His main point is that marginal productivity value under full employment conditions of the agrarian population falls below the ruling wage for agricultural wage-labour, hence the economic viability of the self-employed peasant economy which is able to make a fuller use of the available labour force than a wage-labour economy. The marginal productivity theory of wages is thus the theoretical instrument which Chayanov in fact used. I sometimes suspect that Chayanov's claim to fame has not only come from his valuable empirical work on peasant demography and family structure, but also from the fact that he is believed to have outsmarted both the economists and the Communists. I have dealt with the economists. As far as the Communists are concerned, I consider Chayanov's findings to have been anticipated for instance by Kautsky, who noted with disapproval in *Die Agrarfrage* that the peasant will work harder than the labourer for the same return.

To come back to the productive relations expected in an agriculture which makes little use of capital, one might say

that the fall of marginal productivity value (under full employment) below the ruling wage is the way in which the con-
tradiction between wage-employment and unmechanized agriculture manifests itself. Once agriculture becomes more
mechanized, the economies of scale to be realized gradually
offset the economic loss of not being able to make full use of
the labour available, and one would then expect profit-minded
landowners to resort to wage-employment. Before this
happens — in a laissez faire situation, i.e. in the absence of
legislation protecting tenants, and in the absence of a political
challenge to the legitimacy of rentier landlords — one would
expect profit-minded landowners to turn into rentier landlords,
thus being able to profit, at least in part, from the increased
labour effort of labourers turned into peasants (cash tenants
or sharecroppers). A number of examples from Southern Spain
and from pre-revolutionary Cuba could be given to show that
landowners really thought about such matters in approximately these terms. Thus, in nineteenth century Cuba, when discussing the forms of labour to be used after the end of slavery,
landowners favoured a system of tenancy:

> by dividing the land into small plots, its cultivation will be
> better; in a bad year, the landlords will save the wages that
> otherwise they would have to pay; as the colonos' interest
> will not be limited by a fixed wage, the colonos will make
> an effort to improve cultivation so that sugar cane yields in-
> crease, since this yield will measure their income (Aldama,
> 1862).[1]

If there are such good reasons for turning labourers into tenants
in a non-mechanized agriculture, why was there such a strong
predominance of casual labourers in Southern Spain and in pre-
revolutionary Cuba? We find two interrelated, non-economic
factors, working against the substitution of tenancy for wage-
labour. One is the self-defeating legislation which provides for
security of tenure and a lower rent level than that the labour
market would produce. One of the forms that competition
among wage-labourers takes is that of asking for land as tenants
to be sure of having work and an income. However, if rents are
prevented from reaching the level they would reach by virtue
of such competition, then the incentive for landowners to
change from wage-labour to tenancy decreases. One belief
underwriting the legislation on protected tenancy is fear of an
agricultural proletariat, and support for a rural middle class
and for the peasantry. Another is the notion of the 'social func-

tion of ownership': legislation protecting tenants is one way of registering a protest against rentier owners, and the rural bourgeoisie itself concurs in this. Thus, in Spain *cultivo directo* legitimizes the existence of large holdings against the anti-rentier, anti-absentee, anti-feudal attacks, just as in India 'personal cultivation' (i.e. with wage labour) serves the same purpose. The defence of the large landholders' class interests comes to depend on the use of wage-labour. To quote from a Minister of Agriculture *under the Franco regime:*

> Let us discriminate justly. He who profits from the best of the produce of the fields and from the cream of its income in the idle leisure of a decadent and rapacious absenteeism does not deserve the same assistance as he who, being an enthusiastic and devoted capitalist, takes to the fields, not only his capital, but also his life and devotion . . . (Cavestany, 1951).

To quote from a programme of the Chilean Christian Democrats: 'All the estates which are not personally farmed by their owners, but rather in indirect form through cash-tenants, share-croppers or any other system of farming by third persons, and which constitute for their owners a simple means for profit or amusement, will be subject to compulsory expropriation' (cited in Delgado, 1965:580). Finally, to quote from the report of a meeting of sugar cane planters in Cuba: 'Workers are one thing and we are quite a different thing. They are workers. We are employers, businessmen, men of enterprise and responsibility' (21 May, 1944), cited in J. and V. Martinez-Alier (1972:91—92). In order to be a 'man of enterprise' you have to use wage-labour. Otherwise you run the risk that someone — perhaps an Eric Wolf — will point out that you are a superfluous rentier, siphoning off much needed surplus from your poor tenants.

Then, if for the stated reasons we have a system of wage-labour in an unmechanized agriculture, the Cuban and Spanish paradox of 'men without land and land without men' is likely to arise, at least seasonally, because of the operation of an institutionally set wage level, by definition greater than zero. An attempt is often made to explain the paradox of men without land and land without men by means of a different economic reasoning: 'landowners benefit more by cultivating the land with the help of this labour than by leasing out their lands. If they leased out their surplus land, the cheap labour supply would appreciably diminish and, for want of competition, the

wages of those who were still available would rise. To avoid
this they choose to leave some of their land uncultivated'
(Dube, 1955:75). This explanation assumes economic rationality on the part of the landowners as a group. The question
of how large a group, and therefore how plausible the explanation, will depend on the existence of restrictions to the geographical mobility of labour. Neither legal nor illegally effective restrictions to mobility exist in any of the countries I am
referring to, not even in the highland haciendas of Peru. If
labourers are mobile, then one would need to assume that
agreement is reached among landowners within the territory
where mobility is allowed.

Pseudo-serfdom in Andean Haciendas

Though I have just quoted from a book on India, I am aware
that India is perhaps one country where economic considerations have been perhaps irrelevant to problems of allocation of
labour. This is because of the operation of caste. Consider for
instance Neale's description of the jajmani system:

> The question of how precisely status was used to organize
> [the village] economy . . . can be . . . explained by the con-
> cepts of reciprocity and redistribution . . . It was by virtue
> of each member of each caste within the village fulfilling his
> or her religiously sanctioned duties that the grain heap was
> to be distributed at harvest time . . . There was scant regard
> for economic rationality in the distribution. Some rough
> approximation to work rendered is indicated in the carpen-
> ters' and blacksmith's shares based upon number and size
> of ploughs . . . but this cannot be said of . . . the washerman's
> and barber's shares . . . (Neale 1957:222—27).[2]

I am not competent to judge the ethnographic accuracy of this
description. And I shall not go into the question of whether
'reciprocity' and 'redistribution' are descriptive or explanatory
categories. I wish only to make the point that in such a system
the economist has little to offer in the way of explanation.
Economic explanations of economic phenomena are to be preferred to sociological explanations, if they are available. Thus,
the distribution of produce between landlord and sharecropper
usually leaves to the latter the equivalent to the imputed value
of his contribution (valued at opportunity cost), and is thus
explained. But in the jajmani system there is no economic explanation of this kind. This is *not* because there is no market,

since the absence of a market does not necessarily preclude
the people themselves from choosing between alternatives according to some system of preferences and weights. The economist does not need 'the measuring rod of money' to explain
an economic system (Rappaport, 1967; Salisbury, 1962). Thus,
when the peasant family's behaviour is explained by saying
that the peasants, in the absence of alternative employment
opportunities, will work up to a point where marginal productivity drops below the ruling wage, the economist is not saying
that the peasant will inside his head convert production in
money terms — whether he counts production in dollars, calor-
ies, or kilogrammes, is of no importance provided he estimates
in some consistent way or other the amounts of output produced, and estimates and gives a value (also in some consistent
way or other) to his labour input. In India, it is because religious principles excluded choice in the employment of labour
and its remuneration, that the economist would be unable to
explain the jajmani system. Similarly, the categories 'reciprocity' and 'redistribution' have also been used to good effect to
describe the pre-hispanic Andean economy and society
(Wachtel, 1971).[3] Although such categories do not perhaps
explain the system, they are nevertheless useful to describe it,
perhaps more so than the categories of conventional economics.

This is not the case with the account of the feudal system by
Witold Kula (1970), whose analysis will serve as an introduction
to the economics of Peruvian highland haciendas to which I
shall immediately turn. Kula himself has some rude remarks to
make on conventional economic analysis, but in fact a large
section of his illuminating book is a mere restatement of the
principle of opportunity cost. The system was based on the
existence of serfs who could not leave the estates. Estate
owners produced for the export market. *If* the land had been
imputed a rent, and *if* the serfs' labour services had been imputed a cost (i.e. the cost of the production foregone from
the land occupied by the serfs), then the Polish feudal estates
would have shown accounting losses. The economist is supposed to be baffled by the stability of such an 'uneconomic'
system. However, the economist is not baffled in the least,
since he has been told that there were institutional restrictions
to the mobility of serfs and to the selling of land. Serfs being
unable to leave the estates, it made sense for the landowners
to continue using this system. They tried to use their monopoly
power, very much as slave owners could do in plantation
America, to increase labour obligations, and make the serfs

D

pay dearly in labour services for their own subsistence plots.

Let us now consider haciendas in highland Peru, where the use of labour and land tenure patterns bear a superficial resemblance to a system of serfdom.[4] The question has become confused and controversial, following A. G. Frank's work (1967), which rightly emphasized that the image of South American agriculture as dominated by closed units, with no contact with markets, and belonging to a landed elite of colonial origin, was wrong. (That Frank was right on this point, does not mean that he was right in his notion that capitalism, and not so much the crises of capitalism, underdevelops. For an implicit critique of the so-called 'dependency' theory as related to Cuban history, see J. and V. Martinez-Alier (1972).

Frank has been attacked by attributing to him the view that profit-mindedness and production for the market were incompatible with non-capitalist arrangement (i.e. non-wage-labour), the attack partly missing the mark because Frank analyzed land tenure and use of labour in Brazil, showing, as Caio Prado had done before him (1960), how sharecropping arrangements were introduced with the aim of lowering costs and increasing profits, thus accounting for at least one of the anomalies. Frank did not, however, study Andean haciendas. His critics, notably Laclau (1971) and Romano (1971) have drawn a very misleading parallel between non-capitalist labour systems such as plantation slavery (or Kula's model of serfdom) and the Andean haciendas' labour system. Restrictions to the geographical mobility of inquilinos in nineteenth century Chile were taken for granted by Laclau, while research on this particular topic shows that inquilinos were free to leave the haciendas (Bauer, 1971), though they would rarely do so, since they were doing better than the landless labourers. Neither Laclau nor Frank addressed himself to the study of debt-peonage, which is the institution which some historians (Macera, 1971) would regard as providing the means for Andean landlords to enjoy monopoly power in the labour market, i.e. the intentional burdening of hacienda peasants with debts beyond their capacity to repay them as a means of tying them to the estate. My impression, from the study of twentieth century records, is that this system was not used in this period. Hacienda records also provide evidence of hacienda peasants being threatened with *expulsion* when they seriously misbehaved.

As a contribution to the research on the functioning of Andean haciendas, I shall now illustrate briefly the process of attempted dispossession of the Peruvian highland peasantry, largely unsuccessful, drawing on the correspondence between

Eulogio Fernandini, owner of mines and a large landowner,
and the manager of his estates, between the years 1915 and
1930.[5] This case is not typical of Peruvian highland haciendas.
In the first place, the Fernandini haciendas were not on the
whole agricultural but pastoral haciendas. Secondly, Fernandini
was a progressive man, who imported rams from Scotland, New
Zealand, etc.; he also hired a few Scottish shepherds (later overseers) who I believe did not come directly from Scotland but
by way of Patagonia. We have so far no other catalogued records of highland haciendas which go back further than 1930,
except for this voluminous and substantial Fernandini
correspondence.[6]

Leaving aside the technical questions about sheep farming —
which are nevertheless at the root of the process I shall sum-
marily describe — I want to draw attention to a few points.
First, Fernandini's own mentality. On one occasion he complained that the trouble with Holy Week was that at least two
days of work were lost in his Lima office. He always discouraged excessive religiosity in his haciendas. His general manager,
who knew this well, told him of the situation in one recently
acquired hacienda in the following terms: 'the men have been
used to drunkenness and disorder, to religious festivals, and
to the predominance of the priests of this region. I have found
myself in the need to go against all these customs in the most
radical way, to start with, by forbidding the holding of Mass,
which took place not less than once every two weeks, and also
by giving up little by little the distribution of rum' (July 6,
1927).

Indians were supposed to work. They were, however, not
supposed to be exploited by non-capitalist or pre-capitalist
methods. Fernandini supported the candidacy to political
office of Manuel Vicente Villarán, a Peruvian pedagogue and
a so-called 'positivist', and he engaged him as his lawyer. He
was opposed, nevertheless, as Villarán himself probably was,
to such crude methods of gaining votes as monetary gifts
(January 1917).[6] He disapproved of giving advances to peasants and tying them down in debt, and he was even opposed
to having a cantina — a 'company store' or rather 'hacienda
store' — in his haciendas, since 'this would limit to some extent, even if we do not want to, their own freedom . . . to dis-
pose as they see fit of the fruits of their work' (September
24, 1921). To a certain degree, he might have been worried by
how the urban radicals of Cerro de Pasco, who sometimes
printed attacks on him in their newspapers, might interpret

the opening of cantinas – perhaps as a means of getting the
peasants into debt and thus into working at less than the going
rates. But his refusal both to advance money to peasants who
demanded such advances and to open cantinas has a ring of
truth about it and links very well with his whole outlook on
life.[7]

He was worried by the loss in income he suffered from the
fact that large sections of his haciendas were occupied by the
shepherds' own sheep and food growing plots.[8] Time and again
he gave instructions to count the number of the shepherds'
own sheep, and to limit their number – the usual limit often
mentioned in this period being about 300 sheep per shepherd.
He was also aware of the fact that shepherds 'could pass sheep
belonging to the neighbouring villages or to outside private
persons as if they were their own sheep, charging themselves a
rent' (March 16, 1922). This was no doubt a widespread practice. Shepherds complained when they had to pay rent to
Fernandini for the pasture used by their own sheep exceeding
the free limit of 300 (August 14, 1929, for instance). At the
same time, this did not prevent them from asking for, and
getting, moderate increases in the low wages they earned for
their work in the haciendas' 'demesne' lands. Confusing behaviour for someone like Fernandini, and indeed for many other
landowners. They knew, however, that it would be difficult to
make the Indians change their ways. Even before 1930 and
the rise of APRA, they were occasionally confronted by walk-
outs and refusals to work on the 'demesne' lands, and also by
occasional invasions from neighbouring communities, like the
community of Yanacachi from which at least sixty drunk
individuals carrying flags and drums, moved by *inconsciencia
y tinterillaje,* invaded one hacienda in 1928, destroying land-
markers (August 2, 12, 1928).[9]

External pressure from neighbouring communities could
take the form of an invasion in such a 'primitive rebel' manner,
and could also simultaneously involve long legal battles fought
on the basis of property titles which could go back, say, to
1555. There was also the internal pressure from those who
asked for higher wages for work on the 'demesne' lands, at the
same time keeping a large part of the haciendas in their possession. Both types of pressure start earlier than 1930, in the
Fernandini haciendas.

Fernandini was keen on making money. When the influenza
epidemic reached the haciendas in 1918 he commented on the
use of brandy on Indians as being an 'uneconomic and counter-

productive' waste (November 21, 1918). But, though he was
profit-minded, he dealt carefully with the Indians, so much so
that his general manager (and brother-in-law) told him, in view
of his lenient attitude towards another encroaching community, that 'there is no Indian around this region who does not
say that they can take any advantage they wish from *taita*
Eulogio, because *taita* Eulogio is afraid of them' (November
12, 1929). I do not think that he was actually afraid of the
Indians (perhaps he became so in 1947, when it appeared that
unions had come to stay), but he undoubtedly found that
Peruvian Indian peasants had strange ways of behaving.

Even a Fernandini, the greatest Peruvian entrepreneur, able
to hold his own in the mining industry against the American
companies, found himself in difficulties in rationalizing his
haciendas. Although he was not in the least a typical Peruvian
landowner, his case is all the more instructive. If even a Fernandini could not make proper workers out of the Indians, it
is not surprising that many 'traditional' landowners did not
even make the effort.

In all, however, Fernandini was relatively successful in his
enterprise. Taking wool production as an index, it grew from
782 quintals in 1916 (April 3, 1916) to 2,811 quintals in
1929 (April 22, 1929). By the 1950s, the Fernandini haciendas
(which occupied about 300,000 hectares) were to produce
some 8,000 quintals of wool — in the 1960s the haciendas
were invaded and the Fernandini family disengaged from the
land by selling it to the government. Part of the growth in production came by the buying of haciendas up to the late 1920s;
part came from the expansion of 'demesne' lands inside the
haciendas; and part came from the higher yields of wool per
sheep and per hectare which could be achieved with improved
stock.

In 1962, Ing. Carlos Peña, who had been with the Fernandinis since the 1920s (he had written the graduation thesis for
Eulogio Fernandini's son, Elias, (March 7, 1927) explained
why, against their will, the Fernandinis had not been able to
introduce a wage-labour system in their haciendas: 'to avoid
the turmoil which would arise from the displacement of men
and animals which would put pressure on the comunidades'
(Peña, 1962). Internal peasant encroachment upon hacienda
'demesne' lands would diminish only at the risk of external
encroachment increasing. There was at the same time an awareness (already in the late 1910s, if not before) that the land
occupied by non-hacienda animals and used for the peasants'

own growing of food produced no income to the Fernandinis, but there was also an awareness that it was difficult, though not impossible, to bring this land into the 'demesne' lands.

The great variety of land tenure and use of labour systems might seem to defy not only theorization but even classification. I have nevertheless attempted to explain some of these arrangements, or social formations. The explanations combine economic and sociological analysis. I have identified wage-labour latifundism in an unmechanized agriculture, the exist-ence of which is difficult to explain at the economic level, and which is best explained at the level of landowners' class consciousness. I have explained sharecropping and small cash-tenancy, using Chayanov's notions and translating them into conventional economic language. I have also noted that there might be a different system by which the produce is shared between landlords and peasants in customary proportions, which bear no relation to the market or imputed values of the respective contributions, even after allowing for the difference in the valuation of work if sold on a market or used for one's own family. This system, as found in India, should not be called sharecropping but the jajmani system. I have also taken some tentative steps towards an explanation of Andean haci-endas. While Kula has shown very clearly how a market, and profit-mindedness, were compatible with a system of serfdom — given the institutional set up, i.e. restrictions to the geographi-cal mobility of serfs, inalienability of land, etc. — his model is not applicable to Andean haciendas. They were not 'feudal' — in Kula's sense — since the land was bought and sold, and since landlords tried to displace peasants or to dispossess them, with-out much success (sometimes using the 'anti-feudal' ideology precisely to support such efforts). The stability of productive relations in the Andes analogous in form to serfdom is to be explained at the socio-political level. In a way, there is an analogy with Genovese's interpretation of slavery (Genovese, 1965), and with my own interpretation of wage-labour lati-fundism, in that the socio-political level appears as decisive in this case. The difference is that in Andean haciendas it was not the landlords' class consciousness and prestige which de-manded the persistence of the system but, on the contrary, it was the resistance on the part of the peasants which made it difficult to change it to a system of wage-labour (or share-cropping, or cash-tenancy), which would have been far more respectable and very possibly more profitable.

Politics

There is of course a great distance between theories about systems of use of labour and theories about the peasants' and labourers' political actions. Very broadly speaking, one assumes that there is a correlation between socio-economic position and political attitudes and actions. As are the relations of production, so, more or less, are the forms of political action (and reaction). Though there might be difficulties in explaining why the APRA (a middle-class party) organized the Peruvian sugar workers' unions, and why the Communist Party organized the Cuban unions, and why Andalusian labourers were predominantly anarco-syndicalist, one is not surprised to find unionized labourers in the sugar plantation of coastal northern Peru, or to know that from 1933 onwards cane labourers were unionized in Cuba, or to know that labourers in Southern Spain were unionized up to 1936. To give a very obvious example, one would not expect unions in nineteenth century Cuban plantations run with slave labour; one would expect, if anything, disorganized slave revolts and 'marronage'. One is not surprised at finding Peruvian peasants invading highland haciendas, blowing horns and carrying flags, since one knows that peasants behave in this peculiar manner in other places, in their pre-political phase. But it may be surprising to learn that hacienda peasants in Peru, in the period 1945—47 and again in the early 1960s, formed so-called unions, *sindicatos,* and signed collective agreements with the landlords. One does not expect unions in haciendas which had a type of labour system so similar in form to serfdom, and the question readily arises, were these 'true' unions? or perhaps, were these 'true' peasants?

Sometimes, some of the men who had formed or attempted to form such unions also took part in the mass invasions of haciendas, since they were also members of the surrounding Indian villages. Apart form the ubiquity of horns and flags, which I find uncanny and which presumably shows a special attitude towards the land, I think that I would question the interpretation which distinguishes sharply between labourers and peasants in their attitude towards obtaining (or recovering) land. In fact, that proletarianized, unemployed labourers demand land should not be surprising since in the light of the previous economic analysis (in the section *Sharecropping and Wage-labour*) I have argued in favour of the superiority of the peasant economy over wage-employment, as an arrangement

more conducive to bringing social or opportunity costs in line
with private costs of labour.

I shall argue, therefore, at the political level, that the assumption that peasants, however politically unmotivated, want land,
while truly proletarianized labourers do not want land, is probably an unwarranted one. If we define truly proletarianized
labourers not by their wishes (assumed or demonstrated) but
by their economic situation, then it is quite clear that there
have been truly proletarianized labourers who have wished
land — such as the Andalusian labourers who clamoured for a
reparto, or to some extent Cuban labourers during the land reform of 1959—60, though not, it seems, the plantation labourers of northern coastal Peru, or the English labourers who re-
belled in 1830, or the *braccianti* of the Po Valley.[10] There is,
then, need for a discussion on the desire for land, and on whether we can find any significant difference between peasants
in a pre-political phase, peasants in a political phase, and
labourers. By peasants in a political phase I mean peasants
who are organized by political parties, or at least who make
use of political slogans such as 'land to the tiller'. By peasants
in a pre-political phase I mean (had I ever met any) peasants
perhaps in highland Peru who would, for instance, sponta-
neously attempt to recover the land on the grounds that it belonged to the Inca king of old, paying no attention whatever
to legal forms and not attempting in any way to gain political
support from the towns.

Before going on to consider the political actions of the Peruvian highland peasantry, I shall make some brief remarks on
the Andalusian case. Hobsbawm thought he had encountered
a 'peasant revolutionary movement of the millenarian sort'
even in Andalusia (Hobsbawm, 1958), but I think he was mistaken, for two reasons. First, the anarco-syndicalist movement,
thought it believed in sudden revolution, believed also in collective bargaining and in strikes over such minutiae as the abolition of piecework (which was in fact the reason for the first
recorded strike in Southern Spain in 1883). Secondly, the
reparto of the land preached by the anarco-syndicalists (i.e.
the taking over of the large estates) was not only based on the
justice of recovering common lands lost in the past; it was not
only based on a vague 'right to land by labour' on the analogy
of the Russian 'labour principle' (Perrie, 1972) — although
Andalusian labourers would have found such a principle perfectly reasonable; it was not only based on the slogan 'land to
the tiller', which the anarco-syndicalists shared with the social-

ists; it was also based on the instrumental view of land as a
means of reducing or eliminating unemployment. At least part
of the drive towards the revolutionary occupation of land and
collectivization in 1936 came from the economistic grievance
over unemployment. Perhaps labourers should really have
found sufficient grounds for land invasions in the superfluity
of landlords, or in the robbery of common lands, or in the belief that the millenium had arrived. The grievance over unemployment provided at least additional stimulus. There is just
not enough evidence to say which was the most important
stimulus, and this is not only because of lack of historical research but because the different motives had probably con-
current relevance. Andalusian labourers shared with Andean
peasants, and indeed with peasants the world over and with
industrial workers (but not tribal groups), the need to take into account the likely reaction of the political authorities.
Occupying land on grounds of unemployment would clearly
be less offensive than declaring that landlords were abolished.
On the other hand, one might argue that the grievance over
unemployment was the basic motive for action, though it came
naturally to be reinforced by an ideology, which would be un-
fair to call utopian, since the Civil War was, after all, a close
run thing.

When one finds Andalusian labourers under the Franco regime still demanding a reparto of the large estates, one is not
surprised by the mildness of the phrasing of such petitions:
'energetic measures are demanded from the Government in
order to mitigate unemployment, by means of public works,
compulsory cultivation of the fields, and even a reparto of the
estates which have given up cultivation' (cited in Martinez-Alier,
1971:104). Their courage is surprising, their prudence under-
standable. They are trying not to become too offensive, and
there is nothing specific to Spain in this. Petitions for land in
Brazil today, from the Church-sponsored rural *sindicatos*
which lead a twilight life, are more often based on the reports
of researchers from the Wisconsin Land Tenure Center or from
the A.I.D. than on the theories of less respectable economists
or even on Papal encyclicals. The search for a common ground
for discussion with the political authorities does not exclude
the simultaneous repudiation of landlords, at the level of
thoughts or even of private words: 'bread should be buried
deep in the earth so that everybody would have to dig it up by
himself' — this is the Andalusian version of the 'labour
principle'.

I want to take up the point about the search for a common ground in looking into the political actions (unions, land invasions) of the Peruvian highland peasantry. These are real peasants, who have had enough strength to keep their plots of land and pasture rights either within or without the large estates. They have of course run the risk of dismissal or encroachment, but landlords have lacked the force to proletarianize the Andean peasantry, or to turn it into a more profitable sharecropping or cash-tenancy peasantry.

The roots of the differences between the land systems of highland Peru, Cuba, and Spain go far back into history. Cuba was a country peopled by immigrants. Southern Spain saw first the creation of large estates by the Castilian conquest of the thirteenth century, and then a slow growth of population and of agricultural activity. The large demographic increase came in the nineteenth and twentieth centuries. There was really no dispossessed peasantry, though research on the effects of the disentailment and sale of Church and common lands in the first half of the nineteenth century might qualify this statement. In the Andes, despite the terrible decrease in the population after the Spanish conquest, there remained a peasantry. The existence of a resident population in the large estates known by a collective noun (e.g. *los maqueños,* in hacienda Maco) would strike Andalusian and Cuban landowners as a most peculiar custom. Even where there is no permanently resident population, as in some of the pastoral haciendas, the difficulty in clearing the estates of Indians and Indians' sheep, not because of land tenure legislation but because of custom and successful resistance on the part of the Indians, would have struck Andalusian and Cuban landowners as strange, as would also, of course, the legislation making communities' land inalienable. So, in the Andes, permanent or seasonal unemployment (the Cuban *tiempo muerto,* as it was dramatically called) is not a strong motive for political action and for the wish for land. It will certainly increase in importance.

The search for a common ground does not show itself in the appeal to the political authorities to 'solve' the crises of unemployment, even by taking such extreme measures as expropriation of the estates, on the Andalusian and Cuban pattern. In the Andes, the search for a common ground shows itself in the legalism of the land invasions, so often remarked upon, and also in some of the grievances put forward by the peasants inside the haciendas: the abolition of 'unpaid' labour services, of 'feudal' obligations, and the demand for increased wages.

Land invasions are never called invasions by the actors, but
'recoveries'. What they mean is not so much getting back the
land stolen from them by the Spaniards (i.e. the whole Peruvian territory), as getting back the land rightfully belonging to
the Indian villages according to colonial titles and which re-
putedly has been stolen by the large estates. Invasions are often
carried out only up to a given boundary, the boundary being
described in rather vague terms in colonial titles. Papers are
important. But perhaps their importance is a sign that peasants
are aware that they will have to deal with the courts on their
own terms, or with officials from the Department of Native
Affairs, or even with the police and troops which will be sent
to dislodge them. That legalism is not as deeply seated as is
sometimes assumed is shown by the cases of forgery of titles,
and also by a trick the peasants play sometimes on landowners
and officials: changing place names, so that descriptions given
in colonial titles become even vaguer than usual.

Though the union movement of 1945—47 has not yet been
studied, the catalogues of hacienda papers so far ready, and a
few monographs on highland haciendas recently published
(Alberti, 1973; Kapsoli, 1971; Tullis, 1970) allow us to stress
a few points. Some of the unions seem to have been brought
into being by the direct influence of miners, but we also have
unions led by overseers, and unions led by simple hacienda
peasants. In one case, one Victor Meza, head of an incipient
union in hacienda Laive and leader of a strike, reappeared
some months later as head of an invading comunidad. In
another case, a group of peasants who had been clamouring
for recognition as a comunidad (with the right to land inciden-
tal to becoming a comunidad), turned into a union in 1946 —
from comunidad de Racco to sindicato de Racco.

The idea of forming unions did not come of course from
the Andes, but ultimately from Europe, through a long chain
of 'culture brokers'. But, though perhaps horns are indigenous
to the Andes, flags, drums, and rockets were also brought in
by 'culture brokers', as were also, more importantly, the property titles and the very idea of property titles. Therefore, I
do not think one is entitled to see land invasions based on genuine or forged colonial property titles as being in some way a
more typical form of peasant action than the unions of 1945—
47. Finally in the pliegos de reclamos submitted by the unions
there is a mixture of demands which makes it difficult to
classify the petitioners, since they asked at the same time for
an increase in wages for the days worked on 'demesne' lands,
for the abolition of 'unpaid' labour services, and for an increase,

or at least no reduction, in the amount of land they could use.

Thus we find in Southern Spain, and to some extent in Cuba, truly proletarianized labourers who demand land (or, rather, both land and work). We find Andean peasants demanding higher wages (or, rather, both higher wages and more land). That Andalusian and Cuban labourers demanded land is important, because this makes them in some way similar to 'peasants'. This demand shows a repudiation of the division of labour in society. The contrast with, say, industrial workers or miners is surely great. Miners or industrial workers might occasionally demand the socialization of mines and factories, but they are very unlikely to demand their splitting up and distribution. The 'peasant' consciousness shown by the demands for an individualistic reparto is, of course, based on a specific level of technological backwardness. But, at the same time, there has been a 'proletarian' consciousness among Andalusian and Cuban workers shown in the demands for land *or work*, and in the demands for the socialization or collectivization of the large estates. In economic terms, that is to say as a measure conducive to the reduction of unemployment by changing the valuation of the available labour, collectivization or socialization is not very different from an individualistic reparto.

In highland Peru we also find an element of 'proletarian' consciousness among the peasantry in the demands for higher wages. In fact, the question of a specific 'peasant' longing for the land has to remain in doubt in highland Peru until we are able to explain the reasons for the demand for land; field work and knowledge of Quechua would be needed. That shepherds should be unwilling to reduce the number of their sheep, in spite of the readiness of landlords to compensate them for their loss, is unlikely to be merely because of the attachment they may feel to their flocks: the prospects for a dispossessed shepherd and his family in the depressed labour market were hardly encouraging. Even nowadays, in the new agrarian units set up by the land reform in the pastoral haciendas, the shepherds refuse to allow the substitution of improved sheep for their own sheep, and the unions which, to the government's surprise, are being set up inside the ex-haciendas demand, as they did in 1945—47, both higher wages and no reduction in pasture land rights.

We can now go back to one of the points that is often made about peasants to differentiate them from labourers and other wage workers. If by 'social class' we mean a group similarly placed in the system of productive relations, with a consciousness of its own interests, and organized to defend such interests, then, it is often asserted, the peasantry is not a class, or is a class of 'low classness' (Shanin, 1971:253). The danger here is that, because of the difficulties in knowing peasants' own thoughts, we might have to infer them from their words or even, at great risk, from their actions. I think it would be wrong to infer from the relatively disconnected nature of land invasions in Peru, or from the lack of permanent peasant organizations, or from the failure of the union movement of 1945—47, that Peruvian peasants have been the fragmented units of Marx's 'sack of potatoes'. Marx was of course referring to peasant *owners* in France, not threatened by large estates. Perhaps Russian peasants after 1917 (Shanin, 1972) or Bolivian peasants after 1952 (Pearse, 1972) would also qualify — peasant unionization in Bolivia *after* 1952, but not at all before, served to segment more than to unify.

To infer thoughts from actions might lead us to assert that Andalusian agricultural labourers are not a social class — since under the Franco regime there have been no unions — or to assert that there was no proper working class in Europe prior to the granting of the right of combination. Probably, the Odria coup of 1948 is to be explained at least in part as a way of dealing with the potential threat of the mass organization of the agrarian poor in the Peruvian Sierra, a threat which would arise from their common situation and class consciousness. We know by now how the evolution of Chilean politics before Frei is to a large extent, at least since the 1930s, explained by the need to deal with the (not yet founded) agrarian unions (Affonso, 1972). Analogous considerations are helpful, in my view, in understanding the repressive nature of the present Brazilian regime. This is in contrast to the view that considers the peasantry to be naturally of 'low classness', and thus more or less incapable of organizing themselves and influencing national politics unless mobilized by outside agitators, 'culture brokers' or revolutionary elites. I have little doubt that, sometimes, peasant movements which were on the point of adopting a political character have been pushed back, so to speak, to a pre-political stage by the landowners' and the authorities' refusal to admit that they were no longer dealing with the less dangerous bandits, etc. For instance, when in 1946 the hacendados around Huancayo (in the Central Sierra of Peru) were expecting coordinated invasions and also the formation of unions, they supported the plans to increase the number of policemen on the pretext, as they said in their con-

fidential correspondence, 'of the increase in livestock thieving'.
One suspects that the true, important fight of the Latin
American peasantry over the last fifty years or so has been
over the issue of union recognition, in which they have had some
success. In trying to assess the importance of *potential* peasant
action there is, of course, the very obvious danger of going beyond the evidence. However, the legislation which in many
countries has forbidden rural unions must be included in the
evidence. Although, perhaps, landowners and authorities are
unduly alarmist, I would think that in general they are much
better informed than social scientists.

Another point is apposite here. A usual way of looking at
peasantries is to consider them as 'part societies with part cul-
tures'. An echo of this approach can be seen in the work of
American and Peruvian sociologists who have used the con-
cepts of 'pluralism' (a way of escaping from 'class') and 'domination', in the sense of the unrestricted power of landlords to
impose their interests, facilitated by the existence of only
'vertical' communication and the absence of 'horizontal' communication (this being a way of escaping from 'class struggle').
Though it is a fact that the Peruvian peasantry has, by and
large, a different culture, and speaks a different language, one
should take this not as a datum but as a problem requiring explanation, after 400 years of culture contact — leaving aside
the question of whether one is in favour of national (creole)
integration in Peru, or in favour of disintegration. My own
tentative view is that the remarkable resilience of Indian cultural life is probably to be explained by the fact that it has
been used as an instrument in the class struggle. This is suggested to me not only by the notion of the 'native communities' with inalienable land, but also by complaints from hacienda managers against 'spoiled Indians'. The 'land problem',
for the hacendados — i.e. increasing 'demesne' lands and, in
general, getting more profits out of the haciendas through more
work from the peasants — was an Indian problem. With this in
mind, the recent change in terminology, from 'Indians' to
'peasants', in the context of the current, most bureaucratic,
land reform carried out by a very powerful government, does
not look very encouraging, from an Indian point of view.

Some Conclusions and Recommendations

Unless the economics of land tenure and use of labour
systems are understood, we shall not be able to understand the

very different and indeed opposite reasons for the fact that
['the landlords'] appropriation of part of the peasants' produce,
and even their political and administrative domination has
generally failed to break the basic features of the peasant/land
relationship' (Shanin, 1972:241). There is no reason for surprise, and there is no need for an appeal to a mysterious peasant 'mode of production' or to the specific virtues of the
peasant economy — instead of conventional economics — in
order to understand that the persistence of sharecropping, and
small cash-tenancy, comes from the fact that both sides benefit from the system, which allows a fuller use of the available
labour (also in terms of effort and quality of work) than the
alternative wage-labour system, which produces a divergence
between the private and opportunity costs of the labour available. Sharecropping and small cash-tenancy are a type of incentive wage. The economic advantages of wage labour only
offset the economic advantages of self-employed peasant
labour when the degree of technical complexity is so far advanced that there is no longer need for many labourers or peasants. This is also true, I think, for sugar cane growing. There
need be no reason for surprise, either, about the persistence of
peasants in a feudal system under which peasants are attached
to the land. Even if the peasants in Peruvian highland haciendas
had been really like serfs, it would be a mistake to think that
a situation where there were no markets for land and labour
cannot be studied by conventional economics. A truly feudal
landlord would not have considered the land occupied by peasants as having an implicit cost, since the idea of dislodging
the peasants would not have entered his head — not because
the arrangement was profitable to him, or not only because of
this, but also because such were the institutional restrictions.
In studying Andean haciendas it is good to keep in mind that
peasants could leave if they wanted to; they could also be displaced, but it was socially difficult to do so. The wrong sociological assumption will lead to applying the wrong economic
model. But there is an appropriate economic model to be
applied, which will probably show that the system was less
profitable to landlords than an alternative system of wage-
labour (or sharecropping or cash-tenancy). The anti-feudal
attack in some ways favoured landlords more than peasants.

The economic analysis tells us whether we should stop at
this level to explain the stability of a system of productive relations or whether we should direct our attention to a different
level. For instance, the economic analysis of slavery in planta-

tion America indicates where to look for the reasons for persistence or for change in the system. Thus, the controversies in nineteenth century Cuba about the profitability of slavery between the spokesmen for the Creole planters (opposed to slavery) and the Spanish merchants and planters (in favour of slavery) allow us to discover the 'white Creole' ideology, provided that their anti-slavery pronouncements are not taken at face value, i.e. by studying whether slavery was really as unprofitable as they liked to say it was, compared to the hypothetical alternative of free wage labour. The white Creole planters developed a coherent ideology, against merchants, against Spanish domination, and also against slavery and in favour of white, free immigrants, despite the fact that slavery was probably a profitable system of labour use.[11] In Gramsci's words: 'The proposition contained in the *Preface to a Contribution to the Critique of Political Economy* to the effect that men acquire consciousness of structural conflicts on the level of ideologies should be considered as an affirmation of epistemological and not simply psychological or moral value'(Gramsci, 1971:365). However, in the case of the Hindu jajmani system, and also in the case of the pre-hispanic Andes, perhaps economic analysis does not help. One can always describe the economic consequences of such systems (who produced what, and who got what), but economic models (apart perhaps from the Polanyi—Dalton—Sahlins model, which in any case is of quite a different kind, and not useful for understanding production decisions), do not help in explaining such systems, not even 'in the last resort', since there is no interplay between productive relations and the socio—politico—cultural level.

In the last few years a considerable number of works have been published purporting to explain theoretically slavery, serfdom, share tenancy, the peasant economy, etc. and so on.[12] Part of the literature on economic anthropology is also relevant in this respect. There is no doubt in my mind that this is the right way, and that efforts should be made to construct economic models (of wage-labour, of sharecropping, of Andean 'pseudo-serfdom', for instance) which will take into account the real institutional frameworks on which such systems operate. Shifting to a non-economic explanation, such as the *cultivo directo* or 'personal cultivation' ideology to explain the uneconomic use of wage-labour, or the Indian resistance in Andean haciendas to explain why there was no change away from 'pseudo-serfdom', should not be seen as a failing, but it should not be premature. Resort to purely legalistic explanations, on the other hand, will not do; thus, if yanaconazgo (a form of sharecropping) in the Peruvian coast decreased after the 1947 legislation, a good explanation ought to go into the ideological reasons for the legislation itself.

A proper theory of such social formations will include an industrial sociology or a sociology of work in the different rural settings, and will also *later* include a political sociology of peasantries. There is need in my view to do research on authority, power and legitimacy in the different systems of land tenure and use of labour, starting not so much from the peasants' attitudes towards the state and revolution, but rather towards their work, systems of remuneration, and so on. For instance, one might think that though a system of sharecropping can be explained as being to the mutual economic advantage of both sides, and as reflecting more a contractual relation than a customary relation, perhaps it provides on the other hand a better base for the development of a system of genuine patronage (as distinct from the 'pantomime' of patronage of which Eric Wolf has written) than, say, a system of wage-labour, or a system of slavery.

From the notion of peasant society as a generic type comes the search for peasant political behaviour. One should rather look for different types of political behaviour corresponding to different types of agrarian class societies. If there is a common factor in this present very limited enquiry it is perhaps that peasants and labourers usually seem to look for a procedural common ground in their dealings with the propertied classes and political authorities — but are peasants different from industrial workers in this? When they act in this way (reparto based on unemployment, and unemployment being deplored by the authorities; land invasions based on property titles, papers being highly regarded by the authorities) one might perhaps argue that this is a reflection of the peasant subaltern position in political life. This would then fit in nicely with the defining political characteristic of the peasantry, what Shanin terms 'the "underdog" position — the domination of peasants by outsiders, and the fact that peasants, as a rule, have been kept at arm's length from the social sources of power' (Shanin, 1971:15). However, more than absolute domination there is usually a give and take (clearly so even in the Peruvian highlands); moreover, other classes of society have also been deprived of political power.

In Peru, behaviour which seems to belong to a prepolitical phase coexists with behaviour which belongs to a political phase,

E

such as unions. The search for a procedural common ground is
apparent even in what looks as pre-political behaviour. Thus,
the invasions will be carried out on the strength of papers
which are, or purport to be, the community's colonial titles,
and will reach only up to a given boundary. The use of flags in
land invasions also belongs to the search for a common ground.
By using Peruvian flags, the invaders are signalling 'don't shoot
at us'. Though they are Indians, they know it is a good idea to
appear as Peruvian patriots. It is more or less like using a white
flag. When peasants take action, they appear to look for the
latest fashions (property titles, or the pliegos de reclamos of
collective bargaining) for the very good reason that they do not
want to enter into violent conflict with the political authorities.
More than being inspired by 'culture brokers', they *use* 'culture brokers'. What they ask for, or push for, depends a lot on
what they think they will be able to get, though they might
from time to time misjudge the situation — a pattern of behaviour common enough in other sectors of society. But the legalism of peasant land invasions tells us nothing about the legitimacy of the system from the peasants' point of view. There
is a danger that we classify ways of peasant behaviour according to the presence or absence of different variables, instead of
looking for the meaning that different types of behaviour have
for them. I have illustrated this by showing how it is not true,
contrary to common assumptions, that peasants want only
land while labourers want only higher wages and assured work,
though it remains difficult to know whether land has a purely
instrumental or a terminal value for peasants in highland Peru,
and for Andalusian and Cuban labourers.

There is much difference between, say, unemployment as a
motive for the wish for a reparto of the land in Southern Spain,
and the 'labour principle' or the superfluity of landlords as the
true, deep, motive, and we should not infer the labourers' and
peasants' true thoughts from their words, let alone their actions.
Consider, for instance, the following letter from a Cuban
labourer to Fidel Castro in June 1960:

> Having been a *guajiro* since I was born, and having a good
> disposition towards work, having been a day-labourer all
> my life, but never having had the opportunity to work a
> piece of land, which is my deepest felt longing . . . I wish
> to know whether we who work for wages have any right to
> get land, for I believe that we who work for wages have
> had so far a rougher deal than those who paid rent (cited
> in J. and V. Martinez-Alier, 1972:193).

Did this labourer really wish to have a piece of land of his own, or did he merely wish higher wages and assured work? Perhaps he himself was not too clear about it. The Cuban land reform of 1959–60 was carried out under the slogan 'land to the tiller' and its initial scope was limited. The majority of wage-labourers would not have benefited from it. Hence petitions such as this one. It is interesting to note that this labourer called himself a *guajiro*, a word which connotes an earlier rustic national tradition (like *caipira* in São Paulo or *pagès* in Catalonia), a rather fraudulent tradition — since Cuba, like São Paulo, has in actual fact never had much of a peasantry. Here again there is a search for a common ground with the authorities. No Cuban labourer in his right mind would claim land as a *guajiro* in 1973, but the mood of the authorities, though not necessarily of the labourers, was quite different in 1959–60, when it might have been expected that the rural humiliation of calling oneself a guajiro would possibly bring forth an urban patronizing gesture. Nevertheless, this labourer felt that he had been more exploited as a wage labourer than he would have been as a rent-paying peasant.

While both the peasants from highland Peruvian haciendas, and Andalusian labourers demand higher wages, the apparent reasons for such demands are at first sight quite different. The latter justify such demands on the grounds of increases in the cost of living, etc., as industrial workers do. The former would also use such arguments, but in their pliegos de reclamos would also add that 'unpaid, feudal' labour services should be abolished and that work-days on 'desmesne' land should be paid at the current rates for wage-labour. The complaints against 'feudal abuses' (terminology which clearly came from the cities) did not prevent them from holding on to the land they used. So, more than to behaviour and to words, we should look into the meaning attached to such behaviour and words. One entertaining hypothesis is that peasants will use words and act roles which they expect will gain them some measure of support from at least a section of the town-dwellers — such roles having been defined by intellectuals and social scientists, whose theories about peasants are absorbed by town-dwellers and political authorities. However, sometimes peasants anticipate social scientists (as in the case of the Peruvian unions of 1945–47), and quite often the authorities know more about peasants than the social scientists. It is nevertheless perplexing in Peru to read petitions from peasants who call themselves feudatarios (a word used in the land reform legislation).

Though I hope this paper will have shown how one should go about studying agrarian class societies (including as a topic of study the ideologies about such societies), I am perfectly aware that this is still a half-baked effort. A summary will be useful, even at the risk of repetition.

It is certainly true that peasants and labourers live in different conditions from the urban population. They work in an unmechanized, or little mechanized agricultural sector, and this has an influence on the organization of work, on patterns of settlement, on the administrative structure, on the setting up of class organizations. However, by emphasizing the specificity of peasants we might lose the insights that could be gained by comparing land invasions with factory occupations, or by comparing, for instance, sharecropping and cash-tenancy with schemes for participation or other incentive systems, and also with attempts at workers' control over their own work rhythms, and so on. I have not tried in this paper to compare agrarian and industrial workers. Rather, I have compared, inside the agrarian sector, some types of peasants with agricultural labourers.

I have first considered the economic difference between wage-labour and sharecropping (or small cash-tenancy), which at first sight seems very clear, but which proves on inspection not to be so clear either from the labourers' or from the landowners' point of view. A free labour market might produce a tendency towards the use of sharecropping (or small cash-tenancy) rather than towards the use of wage-labour. Of course, if landowners had no political power they might have difficulty in maintaining their claim to land and therefore in getting profits or rents out of the labourers' or peasants' work. But this criterion, i.e. superior political power, applies equally to landowners who use wage-labour and to landowners who use tenant-labour, and therefore it is not useful to distinguish between labourers and peasants.

I have then considered the land tenure and use of labour system in Andean haciendas. Here again, a definition of peasants which would emphasize their inferior political position and therefore the landlords' ability to extract rents in labour, would miss the main characteristics of the system. Hacienda peasants were free to leave the estates. The analogy with 'serfdom' is then not appropriate. Though it appears that landowners would have liked to dismiss hacienda peasants, or to reduce their holdings, and to substitute cheaper wage-labour (or sharecropping or cash-tenancy) for the more expensive

labour services from hacienda peasants, landowners had great
difficulty in making this change.

Landowners who would like, for economic reasons, to substitute sharecropping for wage-labour might be prevented from
doing so by their awareness that this may make them susceptible to attack as superfluous, rent-receiving, landlords. Peruvian
landowners who would have liked to introduce a wage-labour
system in pastoral haciendas were prevented to some extent
from doing so by peasant resistance, and by their own awareness of the likely peasant reaction. Thus, as a general principle,
the study of such systems deals with the interplay between
social relations of production and class consciousness. The
appropriate way of analysing productive relations is by the use
of conventional economics, with all the complexities required,
such as introducing risk and uncertainty, introducing also
different levels of effort and quality of work, etc. As Witold
Kula's work shows (a little bit *malgré lui*) conventional economics is even useful for studying the feudal system, though
we should be aware of the institutional restrictions obtaining.

The adoption of conventional economic analysis to study
landowners' and peasants' decisions on use of labour does not
of course imply acceptance of the marginal productivity theory
as a theory of distribution of production. Another caveat is
still in order. There are systems of land tenure and use of labour
(such as the Hindu jajmani system) where conventional economic analysis does not appear to be useful, because economic
motivation and choice play no role. In other systems, such as
plantation slavery, sharecropping, wage-labour, serfdom, and
the 'pseudo-serfdom' of Andean haciendas, economic motivation plays a role, but not necessarily a decisive role — it is by
analyzing the economic reasons for stability and change in the
productive relations that our attention is drawn towards non-
economic factors, which might be the ones which in fact explain the stability and change in such social formations, such as
peasant resistance, or landowners' entrepreneurial class
consciousness.

To turn now to politics: apart from saying that there will
be class conflict wherever there are classes (and this is also true
of towns), it is clear that the study of slave revolts, for instance,
will produce generalisation different from the study of peasant
revolts. But the variety of rural social formations makes it
difficult to accept that anything very specific can be said about
peasant politics in general. Perhaps more in a destructive than
a constructive spirit, I have pointed out that it is difficult to

distinguish between specific peasant grievances and political
actions and specific proletarian grievances and political actions.
The apparent paradox here is the Andalusian and Cuban wage-
labourers who demanded land — a peasant-like demand, and
the Peruvian highland peasants who formed unions and demanded higher wages — a proletarian-like demand. Of course,
we could say that the Andalusian and Cuban wage-labourers
demanded land as a way to eliminate unemployment, and this
is after all a proletarian attitude. We could also say that the
reason often adduced by Peruvian highland peasants in asking
for higher wages is that 'unpaid' labour services were 'feudal'
abuses, and that therefore they saw themselves as exploited
peasants paying rent in labour: and we could add that the
unions they formed and tried to form were fake unions, an
idea they got from the towns, not a genuine peasant idea such
as invading land. However, there is no evidence to say whether
the Andalusian and Cuban demand for land came only from
the proletarian (and respectable) grievance over unemployment,
or whether it came also from their peasant-like belief in the
'labour principle' or in the superfluity of landlords. (There
might also be industrial workers who believe in the superfluity
of industrial capitalists.) Analogously, there is no evidence to
conclude that Peruvian highland peasants feel more strongly
about keeping land than about increased wages. That they
used ideas and institutions which came from the towns, such
as property titles, collective bargaining etc., really tells us very
little about their true thoughts, since peasants and labourers,
just as industrial workers, look for a procedural common
ground in their dealings with the propertied classes and with
the political authorities. This common ground is variously su-
plied by provincial lawyers, government officials, political agitators, and other 'culture brokers', who are used by the peasants and labourers. Some of the commonly assumed differences between peasants and labourers have been shown not
to exist at all clearly, in the countries in question and in the
period under consideration.

NOTES

1. These small tenant farmers were called colonos, not to be confused,
however, with the Cuban colonos of later days, some of whom were substantial tenant farmers or landowners who sold cane to the sugar mills.

Much less should they be confused with the colonos of Peruvian highland haciendas, who are service tenants. The question of risk transference
to which the quotation alludes is the cornerstone of Cheung's analysis
(1969). On the other hand, Cheung considers hours or days of work as
homogeneous units of labour, while my point is that sharecropping (and
small cash-tenancy) are incentive wages. I have discussed this in my book
on Southern Spain (1971: especially 33–34, on the relation between
labour supply and, not only the wage level, but the system of remuneration).

2.   T. Scarlett Epstein (1967) has tried to find some 'economic rationality' but, in my view, she merely describes *how* production is shared
rather than *why*. The literature on rural India is of course very large.
For a recent discussion see Stokes (1973).

3.   Wachtel acknowledges his debt to John Murra's work.

4.   The next few pages make points dealt with in greater detail in the
following essay.

5.   The recovery and cataloguing of papers from expropriated haciendas
is being carried out by the Centro de Documentación Agraria, Lima.
This project has institutional support from the Tribunal Agrario, the
University of San Marcos (through Pablo Macera's Seminar), and from
other Peruvian institutions. Financial support comes mostly from the
Joint Committee on Latin American Studies of the American SSRC and
and Council of Learned Societies. The idea of collecting ex-hacienda
papers was born at the Instituto de Estudios Peruanos. I am very indebted
to all these institutions – and in particular to the President of the
Tribunal Agrario, Dr. Guillermo Figallo. The Fernandini correspondence
has been catalogued by Beatriz Madalengoitia, Israel Terry, Humberto
Rodriguez and, last and also least, by myself.

6.   Of course, bribes had to be paid to witnesses and even to judges in
court cases over property titles or over the stealing of livestock; 'justice
sells itself scandalously' (January 24, 1922). There is no sign in this
private correspondence of any rejoicing over such an uncivilized state
of affairs.

7.   He was susceptible to criticism in local newspapers, such as *El Eco de
Junín* (which in October 1919 printed a complaint from the peasants of
hacienda Andachaca), and *La Voz del Cerro* (which in January 1923
printed a complaint from the comunidad de Ticlacoyán against animals
trespassing from the Fernandini haciendas). One of the leaders of a miners'
strike in the neighbouring Chuquitambo Gold Co. was said to have been
a journalist of a little local newspaper (October 10, 1921).

8.   Since shepherds and others who used hacienda land were believed by

landowners to be under labour-service obligations, one can well understand the reluctance to spend cash by employing the wage-labour of outside *maquipureros* (August 22, 1915). One understands also the landowners' wish to force peasants to fulfil their labour obligations. There is
one case in which court proceedings were taken against Fernandini's
manager because he had forced a shepherd's son to take care of one of
the hacienda's flocks, on the justified grounds that his father had 1,000
sheep, 200 llamas, 11 cattle, 6 horses and 4 donkeys on hacienda lands,
and was paying no rent at all. The case, however, did not go well for the
shepherd since he finally left the hacienda with all the animals
(December 14, 1920, February 9, 1921). Shepherds and others were
given land for food growing and for their own animals; this was often
used as an explicit reason not to increase the low wages they earned for
their labour services (February 10, 1922, for instance). This, is, then
very clear evidence that the concession of non-'demesne' land was always seen, or at least came to be seen, as a cost to the hacienda. Thus, a
petition for higher wages for shearing was rejected since this service
should really be done without payment (April 30, 1919). Fernandini
did not, however, favour the use of force to recruit labour for his
'demesne' lands. He opposed the use of the mechanism of *reposiciones*
which in some haciendas was used to keep shepherds on the job. The
losses in hacienda flocks which could not be accounted for by normal
death causes or by accidents (such as foxes) had to be made good by the
shepherd, sometimes in money, sometimes in his own sheep (even at the
rate of two for one). Repayment could be waived if he agreed to work
full-time for the hacienda for a further period of work. This method
was not used in the Fernandini haciendas. There is one case where a
Fernandini employee persecuted a shepherd who fled to the village,
flogged him, and took cattle away from him for *reposición;* the shepherd, it was alleged, was taken ill with pneumonia as a result. The sub-
prefect made the hacienda give back the cattle to the embarrassment of
both the manager and Fernandini (June 1921, 1922). Physical ill-
treatment of Indians was persistently discouraged by Fernandini and his
general manager; there occurred a fair number of cases, however, not least
at the hands of the few Scottish overseers — oblivious of their own history.

Apart from the early awareness that the lands used by peasants implied a cost to the haciendas in foregone production, there are also
references to the mixing of hacienda sheep and *huacchillas* (the shepherds' own sheep) and to the need to fence in the pastures where the
hacienda ewes were to be kept. It was not only the deterioration of the
improved hacienda stock which was in question, but also the irregularity
in lambing, and hence the high mortality rate since lambs were not born
in the appropriate season (January 11, 1925).

9. A *tinterillo* (from inkpot) means a provincial unemployed lawyer
or pseudo-lawyer, similar, I think, to the *paglietta* of Southern Italy.
The boundaries set out in the property titles were often vague, or were
at least thought to be vague. They could also be most inconvenient. Thus,

the community Vico believed its title to include a circle with a radius of
6,000 Castilian *varas* counted from the centre of the village square
(November 26, 1916). Communities became well versed in the subtleties
of litigation. When the invading community of Yanacachi went back to
their village to celebrate Christmas, Fernandini's manager took the opportunity to burn the huts the community had built on the invaded land.
The aim of building huts was to show to the administrative and judicial
authorities that possession was of long date; witnesses would be found
to testify on the antiquity of the settlement. Fernandini himself disapproved of burning the huts: 'I hope this will not bring new complica-
tions, as it might have happened had the [provincial] authorities learnt
of your plans' (December 31, 1930). This was already in 1930. But even
before, the provincial authorities and the Lima authorities were forces
to be reckoned with by both sides. Even a Fernandini (or perhaps one
should say, especially a Fernandini), could be subject to questioning by
the Patronato de la Raza Indígena, and found himself in the need to explain their position to them (October 20, 1928).

10.   Hobsbawm and Rudé (1969) on the revolt of 1830. On Italy, Petri
(1955), Procacci (1972), Zangheri (1960). I am not totally convinced
that the English labourers or the Italian *braccianti* had no wish at all for
land. On England, I find this wish mentioned in J. L. and B. Hammond
(1919:159), though not in Hobsbawm and Rudé. In Italy, it is implicit
in the setting up of collective tenancies managed by the unions in the
Emilia, a perfectly plausible experiment which would be more difficult
in industry. The labourers' resistance in the Po Valley to engaging in
sharecropping arrangements (also found, at the level of desires more
than actions, in Andalusia) arose, not from lack of interest in land, but
from the intention to share the available work (like the demand for the
abolition of piece-work).

11.   This is explained in the following works: Moreno Fraginals (1964),
Pérez de la Riva (1970), Ibarra (1967), and V. Martinez-Alier (1974). It
is *not* explained in Hugh Thomas (1971), where Saco's economic arguments based on the seasonality of work in the sugar industry and the
need to maintain slaves the whole year round are accepted at their face
value, despite the fact that Moreno Fraginals had already disposed of
them by showing the very long hours slaves were forced to work during
the sugar milling season.

12.   Apart from the well-known works, some already cited, by Chayanov
(1966), Witold Kula (1970), Genovese (1965), Conrad and Meyer (1958),
etc., cf. also Domar (1970), Best (1968), North and Thomas (1971),
Bardhan and Srinivasan (1971). The best economic analysis on Andean
haciendas is Schejtman's unpublished thesis (Oxford, 1971). Both
Schejtman and myself have been influenced by Baraona's report on
Ecuador (C.I.D.A. Ecuador, 1965) and by his article in Delgado (1965).
I have profited by reading a still unpublished paper by Shane Hunt,
'The Economics of Haciendas and Plantations in Latin America'; one

hopes the published version will bring into the analysis the fact that hacienda peasants probably did economically better than some people outside (this is why they should be *threatened* with expulsion), and also the fact that the quality of work, and effort in work, depend upon the system of remuneration: three labour supply schedules are needed in his diagrams, each one further to the right, according to whether work is carried out by service-tenants, wage-labourers, or sharecroppers (or cash-tenants).

III

Relations of Production in Highland Peru

This paper presents some tentative conclusions based on the archives of a few expropriated haciendas. The materials so far collected in Lima filled five or six lorries, and would provide enough for several books to be written by several people. The majority of the new archives consist of account books; a much smaller and sometimes more fruitful part consists of correspondence between hacienda managers and landowners, or hacienda managers and communidad authorities. All this material belongs almost totally to the twentieth century and either has been or is in the process of being catalogued. One would need a great digestive capacity to get through it all and a stomach as strong as that of a llama, since the diet includes massacred, beaten-up and black-listed Indians. One of my tentative conclusions, however, is that the Indians' resistance to the expansion of the haciendas made landowners behave at least carefully, if not kindly.

In order to assess properly the value of the records collected in Lima we should consider that haciendas which kept many records are surely not typical, modal haciendas, but are nevertheless crucial to our understanding of attempts at rationalization and the resistance to it. One might even argue that some landowners were able to expand their haciendas and rationalize the labour system only because other more traditional and wiser haciendas did not follow suit in this dangerous disruption of the rules of the game.

External and Internal Encroachment

By 'expansion of the haciendas' I mean the effective exercise of the rights of property inside hacienda boundaries. I do not mean either the shifting of hacienda limits beyond those established in the original land grants by the Crown, nor the crea-

tion of new haciendas on comunidad land ('capturing' comunidades, as this has been called). My impression is that few haciendas were founded after the eighteenth century. There is now
an easy method to confirm this, as well as the frequency of
changes in ownership, by studying the summaries of property
titles included in the *expediente de afectación* for each expropriated hacienda. Many of the haciendas, however, existed
only on paper and then landowners were not able to make peasants pay anything at all for the use of hacienda land. The
history of haciendas is therefore the history of how landowners
attempted to get something out of the Indians who were
occupying hacienda lands. In some cases (notably, sheep haciendas) twentieth century landowners went to the extreme of
trying to dislodge the Indians (as pointed out by Chevalier,
1966).

Though I have studied the Central more than the Southern
Sierra, I suspect that the fantastic increase shown in Chevalier's
article in the number of haciendas in Puno as wool prices went
up in the early decades of this century reveals only the fact
that haciendas which had long existed on paper became proper
haciendas, i.e. haciendas able to get something out of the
Indians or even to dislodge some of them. The increase does
not necessarily indicate that new property titles were granted
or forged. In the Central Sierra, my impression is that all
haciendas had some sort of titles going back to colonial times.

However, such titles have been disputed, and what from the
landowner's point of view might appear as the effective occupation of his own hacienda lands, from the Indians' point of
view might appear as encroachment. Titles have been in doubt
not only, and not mainly, because of their lack of precision.
There has been time enough since colonial times to conduct
proper land surveys. It would be difficult to improve upon the
following account of such disputes, which deserves full tran-
scription:

> A problem of importance and of much worry which
> originated back in the days when the Cerro de Pasco pur-
> chased the haciendas, is the matter of boundary and
> land disputes. Seldom does a week go by without a
> boundary controversy. It might be livestock passing onto
> our grounds or a 'comunidad' claiming parts of our land
> as their property. When the Corporation purchased all
> the haciendas back in 1924 to 1926, they received the

property from the original owners with the boundaries
status quo. Some of the boundaries were natural terrain
features such as rivers or mountain ridges, but none were
marked with any concrete pyramids as permanent and
defined boundary markers.

Several years after the purchase of the haciendas,
boundary disputes arose and many were settled through
the intervention of a Government arbitrator. When they
were settled, the Government issued a decree legalizing
the boundary and the Ganadera constructed concrete
markers along the agreed boundary. Many sections of
our outside boundaries are still without concrete markers
and most of the adjacent 'comunidades' and private
owners of these unmarked sections, continue to claim
that their land overlaps beyond the boundary we are
defending.

These land claims are made through the Dirección de
Asuntos Indígenas, Ministerio de Justicia. Both parties
concerned are called to Lima to present our titles before
an arbitrator and try to reach a solution. It is a regular
occurrence that the communities involved do not recognize our titles to the land, and claim that we have taken
the land away from them. And then the arbitrator tries
to get us to give a little strip of our land to the 'comunidad' in order to solve the situation peacefully. Naturally
we object and state we are the rightful owners and will
not cede one square metre. The meeting adjourns and
the problem is prolonged until a later date. In the meantime the 'comunidad' keeps pushing its livestock onto
this disputed land, and often gets the Senators and
Diputados of their Departamentos and Provincias to put
pressure on the Government to decide in favour of the
'comunidad'.

These disagreements last for one to ten years, and
during this time our boundary riders are constantly
fighting to protect our land. On several occasions our
boundary riders have been cursed, clubbed and hit with
rocks by our neighbours. That is when we have to back
our men, and go out personally to show our support. It
is not very comforting to face and argue with 50 to 200
half drunken members of an opposing 'comunidad' while
standing on the land they claim as theirs. I ceded a small
strip under the circumstances once, and I regret it to

this day, but I learned to withhold my strongest ob-
jections until they are much fewer in numbers and usually
in my office or some other safe place.

We must continue to protect our land and get every
foot of our boundary clearly marked with concrete
pillars, in order to build wire fences for better protection.

There are two weaknesses that make our battles long
and hard. One is the Government's lack of firm support
to the rightful private landowners who occupy the disputed land. Whenever the 'comunidad' claims a piece of
our land, we have to prove that we own it, yet in many
cases they do not have even a piece of paper that re-
sembles a legal title. Other times they have forged titles
which they use to force the issue. The Government
agency gives the 'comunidades' moral support, encourage-
ment and confidence in these affairs, instead of repri-
manding them for their unauthorized, unorthodox and
illegal procedures.

The other weakness is the poor description or method
of designating the boundaries of our titles, which leaves
much in doubt. In some instances, a title will describe
our north boundary as such and such 'comunidad' but
does not name any reference points between us. In other
cases the titles will name mountains or areas, but over
the years and generations, the mountains or areas have
assumed other names and the original names have long
passed from the minds of the old timers. In these cases
we have to depend upon the 'squatter's right' or physical
possession to keep us put. In two early cases, the
'comunidades' took possession of our land and pushed
us off. We have lawsuits pending for years on these two
cases, yet no verdict has been given.

At present we have 15 land disputes pending. The
one with the Community of Tusi has been going on
since 1914, and it is still a perennial headache. They
occupy about 5000 acres of hacienda Paria, our good
land, and are still trying to take more. We cannot
afford to relax along our boundaries or else we will be
taken.

We will soon start fencing our outside boundaries
which have been settled by Government decree and
marked with concrete pyramids. We trust that we will
be able to settle all boundary disputes in the near
future and put up fences as soon as they are settled. We

> can anticipate further trouble after the boundaries are
> fenced, such as wire cutting and fence stealing, but we
> will be ready to meet it when the time comes.[1]

This was written in the peaceful Odría period (1948–56).
Actually, when the time of troubles came around 1960, the
lands of the Corporation, like those of many other landowners,
were invaded. So far, we have only a highly coloured account
of such events, provided by the novelist Manuel Scorza in
Redoble por Rancas and *Historia de Garabombo el invisible*.
A true history might prove even more sensational than Scorza's
contrived epic, in which the peasants, though able to talk with
their horses, appear as very primitive, politically inarticulate
rebels. Apart from the external encroachment of comunidades
upon hacienda lands, there is also the more interesting encroachment within them, for which Baraona coined the words
asedio externo and *asedio interno* (C.I.D.A. Ecuador, 1965).
In 1955, in the haciendas of the Cerro de Pasco Corporation
(in all, some 300,000 hectares) there were 52,000 sheep (or
rather, sheep plus some horses and cattle reduced to sheep
units) which belonged to workers and shepherds. This was
17.33 percent of the total number of sheep units. In less effic-
iently run haciendas, it would be difficult to discover the ratio
at all since they surely did not bother to keep a control on
numbers (Calle, 1968:34–35). In 1955 the Cerro de Pasco
Corporation started a further campaign to reduce the number
of huaccha sheep, as they are called, by setting an upper limit
of 250 sheep units per worker or shepherd, and also by increasing wages in inverse proportion to the number of huaccha
sheep. This system had already been proposed by Peruvian
landowners, though not implemented with any great success.
From the landowners' point of view, having so many alien
sheep in their own haciendas was a form of internal encroachment. From the comunidades' point of view, the prospect of
having these sheep thrown onto their already overcrowded
pasture lands was disturbing. Landowners were well aware of
this, and they seem to have refrained from pushing huaccha
livestock off the haciendas, knowing that this would give
comunidades additional reasons for invading. Displaced Indians,
whether formally members of outside comunidades or not (in
the Central Sierra shepherds were, as a rule, members), would
have to fall back on comunidad resources. Internal encroachment could be reduced only at the risk of increasing external
encroachment.

Restrictions on the Geographical Mobility of Labour

There is one question which needs clearing up, before going on to examine the efforts that were made to change the labour system, and also before considering how the Indians resisted such attempts. This is the fundamental question of restrictions on geographical mobility of labour. If *huacchilleros* and colonos had been unable to leave the estates, then landowners could have used their monopoly power, very much as slave owners could do in plantation America, and made them pay dearly in labour services for their own subsistence. But if they could freely leave the estates, then it would be wrong to talk of 'monopsony' of 'oligopsony' in the labour market. Even where the haciendas were very large, land was not 'monopolized' — there were at least several hundred large landowners in Peru.

As it was, the landowners got the worst of three possible worlds. Huacchilleros and colonos were not serfs or wage-labourers (or sharecroppers or cash-tenants). They had cheap access to hacienda resources and at the same time could not be prevented from leaving the estates. I think that Baraona's remark on huasipungueros in Ecuador applies generally: 'free peons' income is inferior to that of the huasipungueros' (C.I.D.A., Ecuador: 151), and therefore the huasipunguero who left the estate to join the ranks of the day labourers would have been unwise. Jan Bazant has found that around 1850 in hacienda Bocas, in San Luis Potosí, Mexico the economic condition of the resident population was superior to that of casual labourers (Bazant, 1972). Bauer (1971) asserts with great vigour that inquilinos in Chile were free to leave had they so wished. A detailed study of one hacienda in the Yanamarca Valley, in the Central Sierra in Peru, concludes that 'the most extreme of these sanctions (against colonos) was expulsion from the hacienda lands' (Alberti, 1973). There are of course references in the literature to the geographical immobility of colonos in recent times, though not many in scholarly literature. For instance, Tullis has claimed that in the Central Sierra of Peru colonos were not free to move. But his authority for this view is Ciro Alegría, winner as Tullis tells us, of the First Latin American Prize Novel Contest held in the United States (Tullis, 1970: 123).

More worthy of consideration is the view put forward for instance by Macera (1968, 1971), and also by Chevalier on Mexico (1952): that in colonial times debt-peonage was used, i.e. the intentional burdening of hacienda labourers with debts

beyond their capacity to repay them as a means of tying them
to the estate. Though I am not qualified to comment on the
colonial period (and there is little information on the Peruvian
Andes in the nineteenth century) some points are worth mentioning. One consideration is that debt-peonage, in the sense
suggested, was at most a second best solution as compared to
serfdom proper. Another consideration is that made by Gibson,
that 'an Indian worker bent on leaving his hacienda could find
occasion to do so despite his indebtedness', and, further, that
it was the Indians themselves who wished to get indebted. 'As
monetary values came to occupy a large role in Indian society
. . . the hacienda offered a regular or irregular income. To
Indians who had lost their land (largely, of course, to
haciendas) the hacienda provided a dwelling and a means of
livelihood . . . the hacienda was an institution of credit, allowing Indians freely to fall behind in their financial obligations
without losing their jobs or incurring punishment' (Gibson,
1964:252—5).

In one sheep hacienda I have studied, a system somewhat
akin to debt-peonage was used into the 1950s. The shepherds
guaranteed repayment on the money advanced to them on becoming shepherds with their own huaccha sheep. This system
was not used in the bigger firms, such as Cerro de Pasco Corporation, or Sociedad Ganadera del Centro, or Fernandini.
But in hacienda Antapongo, once the initial expense of an advance had been made and a contract for one year signed with
the prospective shepherd who was to take care of hacienda
sheep, the landowner, perhaps making a virtue out of necessity, would attempt to get services for his money and to reduce
labour turnover by retaining shepherds who were indebted,
not allowing them to leave. If they left, some of their own
huaccha sheep would be embargoed, either with or without
the surrounding comunidad authorities' cooperation. If, against
the rules of the hacienda, they left with their own huaccha
sheep, the cooperation of the Guardia Civil for the detention
of the *prófugo* would be sought, sometimes successfully and
sometimes not. Care was taken in Antapongo that debts did
not grow larger than a few weeks' wages, and always much below the value of the debtor's own livestock. I know of no case
where a job was inherited because of debt, though there are
cases of widows' debts being not only tacitly but also explicitly waived. The word *prófugo* appears sometimes also in the
records of other haciendas, used for shepherds who left before
their period of contract was over, but no steps were taken to

F

make them come back. The situation might have been different in other haciendas, and in earlier times. Indeed, assuming
that haciendas were in difficulties in recruiting labour to work
on the hacienda's own ('demesne') lands because the Indians
neither within nor without wanted to work, and assuming
that large advances had therefore to be made (large in relation
to wages), then of course debts might have been widespread.
One may find statements by landowners to the effect that it
was a good idea for the shepherds to be a little in debt so that
they would not leave for the mines. This sort of statement
was made at the beginning of this century for instance by the
manager of hacienda Consac, of Sociedad Ganadera Junín, one
of the haciendas later bought by the Cerro de Pasco Corporation. In my view, it was equivalent to saying that in order to
keep people on the job you have to pay them the going rate,
either as a lump sum in the form of an advance, or little by
little as wages. Of course, once an Indian got into debt the
landowner like any other creditor, did not like the debt to be
defaulted; and sometimes they were successfully defaulted.

What the coastal haciendas or the mines did through *enganche* (i.e. assuring the permanence of Sierra minifundists in
the new proletarian jobs), the Sierra haciendas did mostly by
allowing huacchilleros and colonos free use of hacienda resources to the extent necessary for their standard of living to
be higher than that of labourers in the labour market. It is
actually misleading to talk of haciendas 'allowing' the use of
hacienda resources since as a rule the landowners had not been
asked for permission. One always has to keep in mind the
basic conflict about the actual legitimacy of haciendas which
manifested itself in the dubious legality of property titles, in
the periodic invasions and in the reluctance of colonos and
huacchilleros to reciprocate for their use of hacienda
resources.

Here we should consider why some landowners came to
think that it would be in their economic interest to change
the labour system. The situation is rather complex because we
have at least two main types of highland haciendas: sheep haciendas and agricultural haciendas, as I shall call them with some
degree of simplification, since two different patterns of land
use are sometimes found on the same hacienda. Their economics differ a great deal since there are considerable economies
of scale in sheep farming. On the other hand, there are different patterns of land tenure and use of labour: free use of hacienda resources in a more or less distant past; use of resources

paid for with money (for instance, the *arrendatarios de pastos*) or in kind (*arrendatarios* who paid *hierbaje:* one sheep per year out of every ten, for example); use of resources paid for with labour service called *faenas;* use of resources plus a small cash wage paid for with labour services — this was the most common system: it was the shepherds working under this system who were called 'huacchilleros', while cultivators were called 'colonos'. The alternative would have been to turn huacchilleros into wage-paid shepherds, and to turn colonos into either wage-labourers or, more profitably for landowners, into sharecroppers or cash tenants.

In the case of sheep farming there would be two reasons for choosing the alternative system. First, given the gap between remuneration and performance, there is a basic lack of incentive to work. Secondly there is the economic superiority of large scale sheep farming (in the sense of lower land and capital unit costs). In the case of agricultural haciendas, there would be only one economic reason for choosing the alternative: the same lack of incentive to work. In the case of sheep haciendas the displacement of huaccha sheep and the turning of huacchilleros into wage-paid shepherds would also allow internal fencing and would perhaps make it profitable (depending on the costs of materials and wages) to dismiss shepherds.

As a background to the study of such alternative systems of labour use, one would need to have figures on the income accruing to shepherds and colonos from the use of hacienda resources to compare it with their wages, in order to show that we are dealing more with peasants than with labourers. Figures on total income (both from the use of hacienda resources and from such meagre wages as they received) must then be compared with the wages earned by pure labourers in order to show that we are dealing with peasants who were doing better than labourers and therefore requiring no extra economic coercion to keep them on the estates. Finally, and looking at it from the landowners' point of view, the cost of the prevailing labour system as compared with the stated alternative must be estimated. The cost arose from the loss of income from the lands occupied by shepherds and colonos but the system did also provide benefits in that lower money wages were paid than in the purely wage-labour market; though it is also the case that these lower wages were accompanied by lower effort — shorter working hours by the colonos, for instance.

Let us look first at some figures on the ratio of 'peasant income' to 'wage income' for shepherds and colonos. These and the figures following are offered more to illustrate the line of reasoning I am following than to prove points, many of which must remain in doubt until further work is done. It is difficult to know how many huaccha sheep any shepherd had. Hacienda records list shepherds and the huaccha sheep nominally belonging to each of them, but it is not certain that they did in fact belong to them. Sometimes, but how often and to what extent we do not know, shepherds managed to make *mishipa* livestock appear as huaccha livestock, i.e. to introduce livestock belonging to people outside as if it were their own, either charging a rent or not (in the case of relatives). Sometimes they are alleged to have operated a joint system of huaccha sheep ownership, perhaps inside families, while taking turns at becoming shepherds. Leaving this aside, one can estimate the number of huaccha sheep units per hacienda shepherd (or worker), at never less than 100. In the 1920s, there had been a free allowance of up to 300 sheep in the Fernandini haciendas. As we have seen, in the Cerro de Pasco Corporation haciendas nearly twenty percent of all sheep units belonged to shepherds and other workers in the mid-1950s, while the number of man-years per thousand sheep units was only a little above two — an average, broadly speaking of one hundred huaccha sheep units per shepherd (or other worker). This was the most efficient of sheep farming firms in Peru.

Accepting then this minimum figure of one hundred huaccha sheep per shepherd, let us briefly see the distribution of a shepherd's income as between 'peasant income' and 'wage income'. I shall use the figures offered by the general manager of the Fernandini haciendas in 1947:[2] ($ = soles)

100 sheep with 2 pounds of wool each at $1.50 per pound	$ 300
50 sheep sold, at $20 each, replaced by births	$1000
Consumption of meat (not included)	
	$1300

This would mean a daily wage of $3.60 per hundred sheep units, and while it might have been an exaggeration to call them 'small capitalists', as the manager did, they were undoubtedly peasants rather than wage-labourers, since the wages they got from the hacienda were around one sol per day.

Looked at from the landowners' viewpoint, not all the returns the shepherds got from their huaccha sheep could be counted as a loss to the hacienda, since the returns came not only from the use of hacienda pastures but also from the capital invested in huaccha sheep. The scales of wage increases in inverse proportion to the number of huaccha sheep, which were proposed and sometimes implemented by some landowners, provide an estimate of the landowners' own view of the costs attributable to huaccha sheep. Thus, the manager of the Fernandini haciendas was ready in 1947 to pay $3.50 per day to those who gave up all their huaccha sheep, and $2.00 to those with up to 100. There is a need for systematic work on this, but I think the figures tally in a general way with those from other haciendas. The implicit cost to the hacienda (not only land rent, as we shall see later) was thus estimated as $1.50 per hundred huaccha sheep a day, while the shepherds were estimated to be making $3.60 per hundred huaccha sheep per day — a sum which included both returns on the capital they had invested in sheep ($2.10) and the waived rent ($1.50). Let us note again that $3.60 is probably a low estimate since the average shepherd was able to keep considerably more than 100 huaccha sheep. On the other hand, however, for huaccha sheep both the average of two pounds of wool production per sheep and the birth rate might be considered on the high side.

Let us now consider briefly one agricultural hacienda, the hacienda Maco, near Tarma, of 2,750 hectares, of which 70 were devoted yearly to potato growing, both on colonos and 'demesne' lands, with a seven year rotation. Potato production was then carried out on a total acreage of 490 hectares of which 6 to 7 were in fallow, the rest of the hacienda being permanent pasture land. In 1946 the Indians had 65 percent of the agricultural land and 30 percent of the pasture land — agricultural land was said to be fifty times as valuable.[3] At some time in the past, colonos had been doing even better: 'Formerly the colonos sowed approximately six times as much as the hacienda, which had an acreage of 10 hectares.'[4] This source goes on: 'As the hacienda has increased its potato acreage to 30 hectares, the colonos acreage has decreased from 60 to 40 hectares.' Thus we have a progressive reduction from 60 hectares in an unspecified past, to 45.5 in 1946, to 40 in 1959. This was a hacienda where one of the first unions was formed and where one of the first real strikes took place in the agitated 1945—47 period. The colonos who led the strike were punished by dismissal from the hacienda.

In hacienda Maco, as in agricultural haciendas in general, once landowners were able to impose their hold and claim their rights, they appear to have used a straight system of labour services (without wages) in exchange for the use of hacienda land. With time, small cash wages began to be paid, and the trend was towards increased use of wages. There were two reasons for this: first, pressure from the colonos themselves, who wanted to keep their plots but also get higher wages; second, the landowners' attempts to wean the peasants away from the lands they occupied by offering them cash wages for the days worked on hacienda ('demesne') lands; however, since from the landowners' point of view the peasants were paid well enough by the use of hacienda resources (they were probably doing better than the landless labourers, the obvious comparison), landowners were naturally reluctant to pay cash wages comparable to those of the outside labour market. The alternative was to increase *faenas* as much as possible. The figures for Maco are interesting because they also come from a rationalizing hacienda. Nevertheless they show the greater importance of 'peasant income' as compared with 'wage income'. It was considered in 1946 that the colonos had to work twelve-days *(tareas)* a month to pay the rent for their plots *(chacras);* i.e. the wages for the twelve work-days were considered to equal twenty percent of the value of the production from their plots, this twenty percent being considered a fair rent. That is, 'peasant income' was taken to be five times as large as 'wage income'.

Our first conclusion then is that 'peasant income' was higher than 'wage income' in the total income of huacchilleros and colonos. A second important consideration is that their total income was probably higher than the wages drawn by landless labourers. Thus, in 1947, when huacchilleros in the Fernandini haciendas (assuming they had only 100 sheep each) were estimated to be making $4.60 per day, every day of the year ($1.00 as daily wage, and $3.60 from their sheep), casual labourers were earning at the most $3.00 per day (which would buy some 15 kilogrammes of potatoes at farm prices). The Prefect's report on hacienda Maco (1946) states that daily cash wages for colonos were about fifty percent of those earned by maquipureros; but cash wages were only twenty percent of total income, as has been stated. Though further work is needed, there are grounds for assuming that huacchilleros and colonos did better than landless labourers. Perhaps the strongest evidence in favour of this conclusion is that threats of dismissal were made against those who seriously misbehaved. In 1949, *chacras* were taken away from the most 'subversive' colonos of hacienda Maco, at the same time that the demands from several people from the neighbouring comunidad of Congas asking to become colonos were refused, since Maco already had too many people on its agricultural land and too many huaccha sheep on its pasture land.[5]

The particularly bellicose colonos of hacienda Chinche also formed a union and went on strike in 1945. One of the union's initial demands was that the landlord be prevented from carrying out his policy of 'systematic evictions'. The *acta de conciliación* which was finally signed allowed the landlord to expel only the colonos who did not perform the stipulated labour services in payment for their potato growing plots (Kapsoli, 1971:38, 41). I might be wrong, but I think that proper serfs would not have been threatened with eviction but rather punished corporally, since expulsion would have been tantamount to setting them free to go to the cities and would anyway have been outside the feudal landlord's set of possible choices.

This same point was made by the owner of Maco himself. Writing to his lawyer, Carlos Rizo Patron, he quoted from a petition which the *maqueños* had addressed to the President of the Republic: 'When we ask for some benefits or complain about the abuses against us, we are thrown into the dungeon and threatened with expulsion.' There was indeed a dungeon in Maco, under the chapel. But then, Ing. Mario Cabello had this contribution to make to the comparative study of serfdom: 'This typically medieval picture, of feudal castles and haughty seigneurs, so impressively described, has been damaged by including, let us say a democratic automobile in a medieval scene. For the threat of expulsion is to the Inquisition-like, medieval procedures very much as a 1942 automobile is to the litters of those time.' He added that one single colono had voluntarily left in the last ten years while the hacienda often received applications for entry which could not be granted because all the land was already occupied. The correspondence received by hacienda Maco provides further evidence of this.[6]

There is no denying that there was physical violence, even in the remark in 1949 by the manager of hacienda Acopalca — later a top-ranking official in the current land reform — that an Indian who answered him back had duly got *su pateadura*.[7] Though I think that the really effective violence was meted out by the police and the army and not so much by the haci-

endas' staff, one will have to explain in due course why violence
had not yet become the monopoly of the state. One might even
eventually find out that there was a pattern of violence to force
colonos and shepherds to work on 'demesne' lands instead of
letting them enjoy and expand peacefully their own plots of
land and pasture rights. Even then, the analogy with serfdom
would be inappropriate in that eviction would have no punishment for a serf. Mario Vázquez overlooked this simple point
in his paper 'Hacienda, peonage, and serfdom in the Peruvian
Andes', where he plainly states: 'Apart from the fully entitled
peones and *colonos,* there exist also the *"yanapacuc", "yana-
pakoj", "satjatas", "puchurunas", "chacrate"* or *"pisante",*
who are individuals who are not members, like *peones,* of the
hacienda, many living under the economic tutelage of their
parents or brothers, and who in theory are not obliged to
labour services but who nevertheless perform labour services
from time to time, *out of fear of being evicted from the hacienda'* (Vázquez, 1961:27).[8]

Such threats of expulsion (drawn from Alberti, Kapsoli,
Vázquez, and from the papers of hacienda Maco) all refer to
agricultural haciendas or (as in Chinche) to mixed haciendas.
In purely pastoral haciendas there can remain even less doubt
as to the lack of restrictions from the above to geographical
mobility, and indeed some landowners, keener to push peasants off than to keep them in, soon learned to turn the attack
on 'feudalism' into a defence of capitalism:

> . . . by the introduction of a modern sheep-farming practice
> it will not only become more difficult to steal sheep, but
> all the ills attendant upon (the present) penning and shep-
> herding system can be removed. At present, it may be said
> that the farmer is at the mercy of the Indian shepherd . . .
> Usually each flock of 500 is grazed by an Indian who also
> pastures on the farm 200–300 sheep of his own and need-
> less to say his own sheep receive more attention than do
> those of the patron . . . The farmer . . . may dip his own
> sheep, but the Indian will not submit to having his inter-
> fered with, so that where there are Indian sheep on a farm
> the complete eradication of scab will be impossible . . . This
> difficulty may be overcome by the introduction of a pad-
> dock system . . . This system will also put an end to the
> unsatisfactory relations between patron and Indians, and
> facilitate the complete emancipation (i.e. dispossession) of
> the latter . . . Under the new system the Indian will be com-

pelled to keep his sheep on his own ground, leaving the patron free to improve and develop his farm to the lasting benefit of the country and himself. He will then be able to pay his Indians a much higher rate of wages, thus lifting him from the state of semi-slavery he now is . . . By the establishment of a paddock system Peru will be converted to a sheep farmers' Utopia.[9]

One finds some evidence that living, and even working on a hacienda has long been a desirable alternative for some people. The beginning of a colono-landlord relationship is described as follows in a book published in the 1920s: 'A family which is too poor, which has no land and even less a home to live in, demands entry into an hacienda.'[10] This is also the implication of the figures cited by Cornblit (1970:25n) from a census at the end of the eighteenth century in the provinces where the mita was enforced. Nearly half the Indians were *indios forasteros*, and one fourth 'lived sheltered' in the haciendas. The *indios forasteros* were either escaping from the mita or had already served in the mines, and were free labourers. Perhaps there was a wide free labour market — cities preceded haciendas, in America — and one may assume that those who sheltered themselves in the haciendas did so because they did not see any better economic alternative either in their communities (if they belonged to one) or in the wage labour market.

There is however the question of why, if there was no restriction on mobility, was there a need to import slaves and coolies to work on the coastal haciendas, and why methods such as the *enganche* were needed to bring Indians down to the coast. One answer would be that Indians did economically better in the haciendas (and also, of course, as minifundists in communities) than they could on the coast at the wages the coastal haciendas were prepared to pay, and we must also take into account the high risks of malaria and other coastal diseases at the time. The economic explanation, however, does not account for the apparent reluctance of the unattached Indians to move. Another answer would be that the framework of social relations in the Sierra was such that, even without restrictions on mobility imposed from above, the Indians (including hacienda Indians) would not leave because they had links of kinship and beliefs which gave meaning to their lives. Here one would have to consider the Sierra economy, from the Indians' point of view, as not only a source of employment and livelihood, but also as providing a godfather for one's

children, a patron saint for one's community, and so on. The
potential economic advantage to be gained by leaving could
not be traded for the loss in non-economic relations. I myself
remain unconvinced by this argument. One remaining argument
is that Indians were such poor workers that it had actually
been cheaper to import coolies.

There is a recent book by Antonio Díaz Martínez which is
mainly a compilation of interviews with colonos in the mid-
1960s. (Among the remarkable features of this book is the information it provides on the pattern of marriages of Ayacucho
landowners to Peace Corps girls.) Here is the author's account
of a visit to the wife of a colono:

> He is at work, but his wife receives us kindly. Three children
> with no shoes play on the ground. The house is a single
> room, twenty-five square metres, with mud walls and a roof
> of cane leaves. A small verandah at the front of the room is
> used as a kitchen and dining room. Food is being cooked in
> a stewpot resting on stones. In a corner of the small patio,
> there is a small yard with three pigs and five hens. She says
> they also have five goats, one cow, and one donkey, which
> are grazing. Both she and her husband speak only Quechua
> and have never been to school. Before, they had a *chacrita,*
> and they grew some wheat, and maize, but now they have
> been thrown out, onto the hillside, and she thinks that their
> field will probably produce nothing, because of lack of
> water. She adds that many young peasants have left for
> the coast, in search of work, and that they themselves do
> not leave because they are not sure of finding a job. (Díaz
> Martínez, 1969:228–9) .

Indeed, with such an assortment of domestic animals and
children and with no knowledge of Spanish, they were wise
not to leave. The point is that this woman thought in terms
of alternative employment possibilities. She was dissociating
the economic aspects of her life from other social and political aspects, in the way that people living in a capitalist system
do, though the labour system can hardly be described as capitalist. We do not know whether Peruvian peasants learnt to
think in terms of alternative employment possibilities only in
the 1960s. I would imagine they learnt much earlier.

The Resistance to Change in the Labour System

The hacienda Maco is the only agricultural Sierra hacienda

so far for which we have good catalogued records in the new
archives. No decisive attempt was made there to change to a
system of wage-labour or (better still, from the landowners'
point of view) to sharecropping or cash-tenancy. But I think
there was a clear perception of the disadvantages of the existing labour system. Thus, to quote from a letter from the owner
to the manager:

> Formerly they worked on alternate weeks, one for the haci-
> enda and one for themselves. According to the collective
> agreement of 8 October, 1947, they now have the obliga-
> tion to work 15 days a month as a minimum, that is, four
> days a week, from Monday to Thursday [and six hours a
> day]. If we made them work on Fridays and Saturdays,
> apart from the fact that we should then pay them the
> Sunday wage which would add 17 percent to labour costs
> [the Sunday wage had to be paid if six days had been
> worked], the following week they will not come to work
> because they will need time to attend to their own activi-
> ties. Organization is thus broken and disorder begins. The
> workers start work when they wish or can and tasks are
> not finished on the expected date, as much because of the
> lack of people [coming to work on the hacienda 'demesne'
> land] as because of the low productivity of those who
> work, since those who come to work lack willingness to
> work because they are working against their own particular
> interest, while the rest are attending to their own crops.[11]

One would normally expect small-scale cultivation rather than
wage-labour in unmechanized agriculture, since the labour of
the peasant family has a lower valuation if put to work for the
family than as wage-labour for the hacienda. This is a relevant
consideration when comparing wage-labour to tenancy. When
comparing labour-service tenancy to either wage-labour or to
sharecropping (or cash-tenancy) the relevant consideration is
that, while in sharecropping or cash-tenancy or piece-work
wage-labour performance is, or can be, linked to remuneration,
labour-service tenancy does just the opposite. The suspicion
may be entertained that a system whereby the higher one's
remuneration (i.e. the more hacienda resources there are at
one's disposal), the less time there is available and the less
need there is to work on hacienda ('demesne') lands, is not a
system conducive to profit-maximization, and it will only be
used, as in a system of serfdom, when there are institutional
restrictions on mobility, and therefore where the production

foregone from the land occupied by the serfs does not represent a cost to the estate.

A system was proposed in Maco whereby money wages increased considerably (up to three times as much) for the workdays worked above the obligatory norm. In 1946, the Prefect's report had considered the alternative of substituting a sharecropping system, rejecting it on the spurious grounds that lack of motivation to work, because of a reduced level of needs, would actually result in decreased production. The allegation is not convincing since the colonos went on strike asking for a big increase in money wages which presumably were not to be hoarded.

So, in the case of agricultural haciendas there is some slight evidence of a move towards a wage-labour system, and none (as yet) has been uncovered of a move towards sharecropping or cash-tenancy.[12] We *know* that such changes did not generally occur in agricultural haciendas, except perhaps in a few isolated areas. But the point is not whether they did actually occur, but rather to understand the reasons why landowners should or should not want such changes. However, let us descend from such hypothetical levels and look at some actual instances, which can be documented, of moves towards a wage-labour system in the sheep haciendas.

I have referred earlier to the attempts made in this direction in both the Fernandini and the Cerro de Pasco Corporation haciendas. Here I shall mention the earliest of such attempts we have on record so far (though I should not be surprised if earlier ones were to be subsequently found). In 1945 the manager of hacienda Laive of Sociedad Ganadera de Centro, Ing. Rigoberto Calle, explained that the reason why wool production per hacienda sheep had decreased in one year from 4.48 to 4.18 pounds had to do with the excessive number of animals in the hacienda: 'When I arrived in Laive, the first question I attempted to cope with was the huaccha problem; precisely, when I tried to explain to the shepherds the new system of increased wages with a discount for huaccha sheep, there was a reaction on their part which ended with a sort of attempt to strike. Seeing this, I had no other alternative but to relent, telling them that they were to work under the same conditions as before.' Another difficulty he saw in reducing the huaccha allowance was that dissatisfied shepherds would go to the neighbouring haciendas, Antapongo and Tucle. To prevent this, he suggested that an agreement among the three haciendas should operate to black-list shepherds who left one hacienda to go and work in one of the other two. It was operated, not always with success. In this way, 'when the neighbouring comunidades overflow with sheep, we shall be able to get good shepherds without difficulty and with a lower number of huaccha sheep than they now have.[13] One is inclined to think that the procedure lacked wisdom; to substituted *asedio externo* for *asedio interno* was really no solution at all. The three haciendas were partly invaded in 1946 despite the fact that no reduction of huaccha sheep had yet taken place. I am not implying that they were invaded because the Indians got wind of such plans. They were invaded because of the changed political situation in Lima. But it was unwise to propose such methods without the force necessary to implement them.

The economic reason why landowners were prepared to pay higher wages to shepherds who gave up their huaccha sheep — even perhaps to the point of increasing the shepherd's total income — was because landowners could make more money out of the pasture land with their own improved sheep than the shepherds could with their inferior ones. If not, they might have considered a system of money rents or of rents in kind, perhaps adapting the institution of *hierbaje*. But wool production per hacienda sheep was more than twice as high, the birth rate was also higher, and so was meat production. Haciendas had imported new breeds, and they were large enough to separate the different sheep by age and kind allocating pastures to them according to technical needs, having lambs all born in the appropriate season, etc. Huaccha sheep, moreover, not only occupied valuable pasture land, but might also interbreed with hacienda sheep, and transmit their diseases to them. Moreover, stealing was made easier by the fact that shepherds had their own sheep which they could sell. Steps were taken to prevent such losses — such as separating huaccha sheep from their owners, compulsory dipping and compulsory castration of huaccha rams. It may eventually be possible to quantify the cost of such measures for the haciendas.

Of course, had it been absolutely impossible to turn shepherds and colonos into wage-labourers (or sharecroppers or cash-tenants) or to dislodge them, then the land they occupied would have had no real cost to the hacienda. Although in practice, they were rarely turned into wage-labourers (or share-croppers or cash-tenants) or dislodged, there was no legal prohibition against doing so — particularly since wage-labour (or even sharecropping or cash-tenancy) were far more respectable than 'serf' labour. Naturally, the legal provisions are not the heart of the matter. What we want to know is whether landowners considered they suffered a loss in income because of the land occupied by shepherds and colonos.

Some — perhaps many — landowners certainly did believe that they had. After all, the system implied that the land occupied by colonos and shepherds was recognized by all parties as belonging in some sense to the landlord. This is why colonos and shepherds were paid either no money wages or money wages inferior to those of outside labourers, since they also received a remuneration *from the hacienda* in the form of use of land. It would be absurd to argue that landowners were not aware of this fact. If they saved in money wages it was because of the loss in income from the land occupied by colonos and shepherds — they could make comparisons with the production from their own ('demesne') lands and with the wages paid in the labour market. If few landowners made attempts to turn colonos and shepherds into wage-labourers (or sharecroppers or cash-tenants), or to dislodge them altogether, this was not, I would strongly argue, because landowners had not noticed that they were prevented from gaining income from the occupied lands, but rather because they felt unable to make such profitable changes, or were even afraid of the Indians' reaction. They had to behave in a paternalistic fashion because they lacked power.

Although, as has been said, colonos and shepherds were getting higher 'peasant incomes' than 'wage incomes', their main grievance appears to have been over their low money wages. The strike in Maco in 1946 was to demand an increase in colonos' daily wage from one sol to four soles. But, although there was undoubtedly a long term trend towards cutting down on their use of hacienda resources, colonos and shepherds also resisted such attempts with some success. I believe that their way of putting up resistance was based in part in the use of cultural differences to further their economic interests. Thus, one finds the manager of hacienda Laive complaining: 'They argue that the traditional custom has been to admit huaccha sheep without charge and that this system should continue, even with the increased wages. And, as you know, the Indian, as a good Indian, is like a piece of granite when he is not willing to understand.'[14] Such a breakdown in vertical communcations proves rather damaging to the theories of W. F. Whyte, F. Lamond Tullis, and other writers of the same school on Peruvian haciendas, though it is fair to remark that such evidence was not available to them (see J. Matos Mar et al., 1969).

I have focused in this paper on the labour system internal to the hacienda, and on some factors which impinged upon the landowners' decisions on land tenure and use of labour. Their field of decision was rather limited. Throughout I have abstained from bringing in such external factors as the demographic situation, or market opportunities, and I have scarcely referred to the general political situations (i.e. the political complexion of Lima governments, which is not a totally exogenous factor since it tends to respond to some extent to the threats from the Sierra). Such factors are clearly relevant in that they might have decreased or increased the attractiveness of a change in the labour system. But I think that, with this qualification, it is profitable to leave such external factors aside so that due emphasis is given to internal factors. The trend towards wage-labour (or, in agricultural haciendas, towards sharecropping or cash tenancy) did not make much progress because of the resistance of the Indians to giving up the hacienda resources they used freely, and because of the landowners' inability to overcome such resistance. In Peru head-on collisions were narrowly averted in 1945—47 and in the early 1960s. In the first case this was avoided probably by the APRA's leaders' refusal to support the unions that had started to flourish on the haciendas, and which Aprista agents had helped to set up; in the second case the state intervened where the landowners had given up. In both cases there was violent intervention by the police and the army.

The novelty of my tentative conclusions, which turn prevailing interpretation upside-down, or perhaps inside-out, results from the fact that the hacienda records now available allow us to look at things from the point of view of the landowning class. Landowners were not even strong enough to have the property titles situation cleared up, nor to have the communities' land brought into the market. More effort is usually devoted to denouncing the sporadic lack of compliance with the legislation which made the communities' land inalienable than to analyzing its significance as an exceptional measure in a capitalist society. One would expect, in revising the debates on this topic, to find reference to the fear of Indian rebellions. The stage is now set for what Macera has called the 'Third Conquest', after the colonial and republican ones. It might again prove to be a partial failure despite the fact that the state commands more power than the landowners did.

Although the new archives of hacienda papers contain many letters from comunidades, petitions from colonos and huacchilleros, etc., one cannot fully understand the situation from the Indians' point of view without learning Quechua and doing field work. One might speculate for instance on the reasons why shepherds refused to cut down the number of their own sheep even though landlords were possibly ready to increase their remuneration up to a level which would more than offset this loss. Perhaps it is because of a genuine attachment Indian shepherds feel for their sheep: the word 'huaccha' also means 'orphan' and 'poor'. Perhaps they correctly thought and still think that despite any short term mutual benefits, in the long run a dispossessed shepherd and his sons and daughters would not fare well in the depressed Peruvian labour market. Landlords (and nowadays land reform authorities and SAIS managers) would also point out with some justification that the pasture land allowance did not benefit the poor shepherds as much as the relatively rich sheep-owners of the internally differentiated surrounding communities. There is also the question of complementarity with other activities at a lower ecological level. For instance, the oxen needed to plough the land of the comunidad of Chongos Alto had access to the pastures of hacienda Antapongo, or could be bought cheaply from the hacienda, provided that the comunero who used the pasture land or bought the oxen agreed to work for the hacienda for a period of time.[15] The records of hacienda Antapongo contain many references to this practice. Lack of access to hacienda pasture would thus produce negative secondary effects on the communities economy, apart from the loss of the direct returns (meat, wool, also dung for fuel, etc.).

Finally, and from the Indians' viewpoint, one would also like to see whether it can be argued, as I think it can, that the remarkable resilience of Indian cultural life comes partly from the fact that their culture has been an instrument against landowners who wanted to rationalize labour systems, to settle boundary questions, to buy comunidad land, etc. This would be in contrast to the view that 'explains' the lack of Indian 'integration' by lack of schools, as if it had been in the Indians' interest to 'integrate', to become 'castilianized'.[16] There remains the question of accounting for other interpretations, which are interesting because of what they say about their proponents. There is, for instance, what I like to call the village shopkeeper's view: Indians are assumed to have been 'feudally' abused and exploited since they were not paid good cash wages to spend in the shops, and in addition they are assumed to have been burdened with debts in the haciendas' cantinas (or *tiendas de raya,* in Mexico). Hacienda unions in Peru in 1945—47 sometimes asked that cantinas be installed in the haciendas so that peasants would have no need to travel to the villages and towns.

By looking at things from the landowners' viewpoint, attention has been drawn in this paper to the actual roles of the Indians as historical agents. A common urban ideology sees the Indians as absolutely crushed under the weight of land-owners' domination, and at the same time proclaiming the Indian as the bearer of national redemption along the lines popularized in Mexico. The inheritor of Inca greatness is ex-pected to step forth eventually onto the stage of history, hope-fully draped in a Peruvian flag. Such millenarian hopes on the part of the ideologues conveniently postponed the precise hour at which the Indian was expected to step forth, and simultane-ously raised the Indian above the sordid materialistic considera-tions of the class struggle.[17] By blaming landowners for wil-fully preventing Indian 'integration', this ideology also had the virtue of eschewing the painful question of whether Peru is or is not, and should or should not be, a nation. The identifica-tion of deplorable backwardness with Quechua speaking comes out clearly in the censuses, where Indians who speak Quechua have also been additionally classified as illiterate, which is rather like calling Puerto Ricans illiterate because they cannot read English. (Anybody would agree that 'illiteracy' is very bad.)

The present military government finally took it upon itself to redeem the Indian, stepping in when the battle was half won. National integration has been achieved by such procedures as renaming 'El día del Indio' as 'El día del Campesino'. Noises have been made about bilingual teaching in the new educational law, but this is not to be taken seriously. To make children bi-lingually literate would require longer schooling and more fin-ancial resources than the Lima authorities are prepared to con-sider. Monolingual literacy in Quechua would, I suppose, be unpatriotic.[18]

The meaning of the current land reform is still ambiguous. In pastoral haciendas, the government's intention is to continue the rationalization process which private landowners had no power to carry through. However, in agricultural haciendas in the Sierra the situation is less clear cut, since many of them are too small and poor to pay for the bureaucrats needed for rationalization. Although attempts are being made to form larger units (as in Pampa de Anta and elsewhere), the government may eventually decide that this is too difficult and dangerous. One cannot help reflecting that a land reform representing different class interests, carried out by a personnel enjoying the Indian peasants' confidence, would not find it so difficult to push through measures which ideally should be to everybody's advantage, such as substituting improved sheep for huaccha sheep.

NOTES

1.   Centro de Documentación Agraria (hereafter C.D.A.). Cerro de Pasco Corporation, Ganadera Division, Interdepartmental Correspondence, from R. H. Wright to J. W. Henley, 5th July, 1952. One should add that sometimes the 'old timers' appear to have changed place names on purpose. Names are of course in Quechua, and it makes sense in Peru to accuse landlords coming from New York, or from Lima for that matter, of not knowing the true names of their own fields. (One example of this in C.D.A., Hacienda Antapongo, Correspondencia Administrador-Contador y Subadministrador, 1946—7.) Apart from forging titles, another indication of the cynical attitude of comunidades towards legal matters is the building of huts immediately upon invasion; the idea is then to find witnesses, paid or unpaid, who will testify on the antiquity of the settlement.

2.   C.D.A. Algolán, Correspondencia Administrador General-Eulogio Fernandini (20 March 1947).

3.   C.D.A. Sociedad Ganadera del Centro, Correspondencia e informes: administración general, 1938—49, Informe (del Prefecto de Junín) al Ministerio de Justicia y Trabajo (20 February 1946) (this report refers to the unions and strikes in haciendas Maco and Queta).

4.   C.D.A., Hacienda Maco, Correspondencia Director-Gerente a la Administración 1946—67, 27 September, 1959.

5.   C.D.A., Hacienda Maco, Correspondencia Administrador al Director-Gerente 1944—69 (26 August 1949).

6.   C.D.A., Hacienda Maco, Documentos varios, décadas 1940—50—50, letter of 8 January 1946, accompanying *El memorial de los maqueños*. Ing. Mario Cabello had written a thesis in 1927 on hacienda Udima, in Lambayeque (there is a copy in Universidad Nacional Agraria, Lima)

where he proposed expanding 'demesne' lands at the expense of colonos, or *pachaqueros* as they are called in Udima. Its part-owner and manager, Ricardo de la Piedra, complained after the land reform of 1969, 'who would possibly have bought this hacienda from us, so full of people and with so little hacienda livestock?' I am grateful to Douglas Horton for this quotation and for information on Udima and on Ing. Cabello's thesis.

7. C.D.A. Sociedad Ganadera del Centro, Hacienda Acopalca, Correspondencia Administrador-Gerente, 1947—64. *Patear* means to kick. There was a court case and the incident was reported in the press, prompting an enquiry from the general manager in Lima.

8. Vázquez was much involved in the famous Vicos project, which no student of Peruvian haciendas can ignore. The absence of any reference to the unions of 1945—47 in Vázquez's work (and also, incidentally, in Hugo Neira's, 1970) is due, to put it kindly, to easily vincible ignorance since the best records of such events (partly used by Kapsoli, 1971) have been kept in the Ministerio de Trabajo y Comunidades (as it is now called). It was unscientific to select for study, on grounds of political expediency and not of historic relevance, only one of the possible types of change in the hacienda system. At least two alternative situations should have been considered: no change (or accurate study of the situation previous to the Cornell takeover), and change through unions.

9. 'Huacauta Sheep Ranch', *West Coast Leader* (March 16, 1926:6—7). (I am grateful to Geoff Bertram for this reference.) A version of this article appeared in Spanish in *La Vida Agrícola* (Lima) (3—4 April 1926).
     Hacienda Huacauta, near Chuquibambilla, in Puno, belonged to Bedoya and Revie company. (I am grateful to Colin Harding for this information.)

10. A. Serruto, *Monografía del distrito de Pichacani*, cited in Espinoza and Malpica (1970:311).

11. C.D.A., Hacienda Maco, Correspondencia Director-Gerente a la Administración 1946—67 (27 September 1959).

12. See however the amusing comment in C.I.D.A., Peru (1966:144) on a small estate in Ancash: 'El conductor . . . más bien desearía convertir a los colonos en arrendatarios y vivir de la renta que obtenga. El encuestador afirma, empero, *por razones que no aparecen claras*, que "los colonos no acceptan ningún cambio en su condición y se aferran a ella".' The comment is amusing because it betrays the writer's assumption that exploitation in the form of payment of rent in labour is greater than under a wage-labour system or under sharecropping or cash-tenancy (this being a wrong assumption, in the absence of restrictions on geo-

graphical mobility). The writer is then perplexed by the resistance on the part of the colonos to change from labour-service tenancy to cash-tenancy.

13. C.D.A. Sociedad Ganadera del Centro, Hacienda Laive, Correspondencia Administrador—Gerente (8 March 1945).

14. C.D.A. Sociedad Ganadera del Centro, Hacienda Laive, Correspondencia Administrador-Gerente (3 February 1945).

15. See letter No. 2 in Martinez-Alier (1973b:113).

16. It should be pointed out, however, that the demand for schools often appears in the lists of colonos' grievances. Perhaps myself, as a Catalan, tend to find 'castilianization' excessively deplorable. It is also true that while the recent change in official terminology (from Indian to 'campesino') does not necessarily mean a distinct improvement for the people in question, and while being called and behaving as an Indian was an advantage in land invasions and in defending a peasant mode of life, on the other hand landlords themselves sometimes found it useful to call them Indians since some forms of struggle were then, in a way, implicitly denied to them. Thus, the manager of hacienda Laive commented on the unions in 1947, 'this is an entirely grotesque thing, because the Indians do not even know how to pronounce the well-known terms such as *comandos, sectores, departamento de capacitación, enlaces*' (C.D.A., Sociedad Ganadera del Centro, Hacienda Laive, Correspondencia Administrador-Gerente, 1940—51 (July 1, 1947).

17. I have used in this paragraph notes by Geoff Bertram on an early draft of this paper.

18. People in Chongos Alto, near Huancayo, told me repeatedly that, while Cuzco Quechua could be written, their own dialect certainly could not, since its sounds were too difficult.

IV

The Cuban Sugar Cane Planters, 1934–1960

Cuban history from 'independence' in 1902 to the mid-1920s
is a history of great economic expansion brought about by
sugar exports and dependence from the United States. Between 1900 and 1925 population more than doubled (from
1½ to 3½ million), and sugar production grew from about 1
million to 5 million tons, going through some pronounced
cycles such as the 'dance of the millions' in 1920 and the
temporary crash in the sugar market in 1921. It would be an
exaggeration to say that the invasion of American capital displaced Cuban or Spanish entrepreneurs. In large areas of the
country (Camagüey, Oriente) there was nobody to be displaced, and there was no other source of finance nor any
market comparable to the United States.

The second phase of the economic history of Republican
Cuba set in with the first restriction on sugar production in
1926. Since then 'restriction' and 'reciprocity' were the key
words of Cuban economic policy up to 1959. Sugar output
was fixed in an effort to keep up prices. This was the beginning of very heavy government interference in the economy.
Because of the buoyant state of the economy up to 1925 and
of the foreign exchange which had accumulated, the level of
economic activity during the late 1920s was kept up through
a programme of public works — it was then that the central
highway was built. President Machado also introduced
temporary protective tariffs in 1927. But at the depth of the
crisis, when Machado was ousted in 1933, sugar production
had dropped below 2 million tons, and the sugar price to a
fraction of 1 cent per pound when a normal price would have
been 3 or 4 cents per pound.

The 1930s were a period of nationalist reaction, the high
point of which was the revolution of 1933. Most Cuban

politicians until 1959 had their political roots in 1933, including Presidents Batista (1940—44, 1952—58), Grau (1933, 1944—48), and Prio (1948—52). The 'authenticity' and 'orthodoxy' of the two main political parties of the 1940s and 1950s, the Autentico and Ortodoxo parties, were based on their loyalty to the nationalist ideals of 1933. Fidel Castro's first speech after Batista had fled the country on December 31, 1958 was on the theme that *this time* the nationalist revolution would not compromise.

Two economic choices confronted the new political leaders in 1933: to substitute imports, or to come to an agreement with the new Roosevelt administration in order to achieve a sugar quota at an acceptable price in the United States' market. This second alternative was the one chosen. A Reciprocity Treaty was signed in 1934 by which Cuba was assured of a share of the American market. 'Reciprocity' meant exporting sugar and sugar by-products in preferential markets in exchange for concessions in the Cuban market. The principle of reciprocity had served the Cuban economy well from 'independence' to 1925, and it extricated the Cuban economy from the Depression in the cheapest possible way — by using the under-utilized capacity of the sugar industry.

The first Reciprocity Treaty with Republican Cuba had been signed in 1903. Hugh Thomas comments that the 1903 Treaty was followed by an upsurge of American investment. 'The investment was almost all in sugar, but whatever remote hope there might have been that the old sugar economy might be replaced by a more balanced system was lost. Root and Roosevelt [Theodore] were giving Cuban mills a chance to expand production till they supplied all the sugar that the U.S. needed. Both politicians, however, believed that they were helping Cuba, not ruining it, and most Cubans agreed at the time' (1971:469). They agreed again, quite rightly, in 1934, and it is wrong to say that emphasis on sugar 'ruined' Cuba either before or after 1934, unless 'ruin' is meant in a moral and not in an economic sense. Everybody may feel entitled to moral judgements and Hugh Thomas' are interesting because they underlie the interpretation of Cuban history which has been predominant until recently and which has been succinctly expressed in the dilemma 'Nation or Plantation'.

This is a neat and suggestive formula but it hides the fact that there have been two conceptions of what the Cuban nation should be. Thus, in the nineteenth century we find the *mambí* independentists setting the slaves free, burning the plantations, and declaring against the racial segregation that the slave plantation system imposed by saying, as Antonio Maceo said in 1870, 'here there are neither whites nor blacks, but only Cubans' (quoted in Ibarra, 1967:52). But, we also find the 'white Creole' conception of Cuban national identity.

The colono system of separating cane cultivation from milling was first proposed by the Count of Pozos Dulces. He had first been an 'annexationist' believing that annexation to the United States in 1851 would remove Spanish tariffs and taxes while preserving slavery. Later on he converted to 'reformism', opposing slavery and hoping that Spain would concede political and economic reforms. In 1857, Pozos Dulces wrote that 'the separation of cane cultivation from sugar manufacture was the best means of solving the terrible question of slavery; white immigrants should be encouraged to plant cane and sell it to the large sugar factories' (quoted in Thomas 1971:275). Pozos Dulces also argued that colonos were to be preferred to wage-labour. Slaves, like proletarians later, spelled trouble. What sorts of trouble, we cannot go into here, but they had to do with the techniques of sugar manufacturing, and with the Spanish policy of supporting slavery and racial segregation in order to be able to threaten the Creole planters with the dilemma that Cuba would be Spanish or 'African'.

More relevant here is to notice the continuity in interpretation of Cuban sugar society from the mature Arango, through Saco, to the 'reformist' Pozos Dulces, and then to Fernando Ortiz and Ramiro Guerra, and finally to Hugh Thomas. The latter explains how sugar had, at the end of the eighteenth century, displaced tobacco as the main commercial agricultural produce, and also dislodged the small tobacco farmers of Güines. Coffee tried in vain to compete with sugar — for slaves, more than for land - during the first four decades of the nineteenth century. Hugh Thomas concludes that 'the tragedy for Cuba in the decline of coffee is that this product could have been developed much more easily by white farmers or small black freeholders than sugar could' (1971:132). While Hugh Thomas contrasts sugar cane with coffee, Fernando Ortiz contrasted 'cane [which] demanded seasonal mass labour [with] tobacco [which demanded] constant attention from a few experienced men. Cane was grown on large estates, tobacco on small holdings . . . Finally, sugar spelled slavery, tobacco freedom' and therefore Ortiz suggested (this is Hugh Thomas' conclusion), 'Cuba should escape from sugar into tobacco production' (1971:1161 and Ortiz, 1946).

If Cuba had been in 1959 something like an immense plantation with, on the one side, a few gigantic sugar mills controll-
ing all the land, under absentee American ownership and
expatriate managers and, on the other, with a large proletariat,
industrial and agricultural, then a nationalist revolution would
have meant a socialist revolution. This was, however, most
definitely *not* the situation. There were no plantations in Cuba.
But, what is a plantation?

If one accepts the definition of a plantation given by Sidney
Mintz as 'a corporate land-and-factory combine' (1959:49),
then there were practically no plantations in Cuba. Certainly
a much lower proportion of the sugar cane was grown in 'corporate land-and-factory combines' in the twentieth century
than in the nineteenth century. However, if one differentiates
between planter-industrialist and planter-farmer, as Hutchinson
does for north-eastern Brazil (1959:38), then there were plantations in Cuba owned by planter-farmers called colonos in Cuba.
Why not call them simply 'farmers'? But did, say, English
large farmers do manual work before agriculture became
mechanized? Cuban *colonias* were really farms, some large,
some small, producing mainly sugar cane, the work being carried out either by the colono himself and his family, or with
the help of hired labour, or by hired labour exclusively. The
sugar cane had to be sold to mills which enjoyed a monopolistic position because of the heavy cost of the transport of cane,
and because the mills owning land made cane supply a condition for continuing the tenancy of the land.[1] This dependence
on the mills made the colonos eager to strengthen their own
position. A national Association of Colonos came into being
in 1934 in the aftermath of the 1933 revolution.

The Colonos' Political Views[2]

Information is particularly good about the views of the large
cane planters who ran the Association. Located as they were
between mill owners and agricultural workers and necessarily
involved in sugar policy making both with the Cuban and
American governments on such matters as sugar quotas and
prices, the large cane planters provide a good vantage point to
understand the structure of the sugar industry and, through
this, the interplay between imperialism and internal class conflict in Cuba. There were nationalists with perhaps as much
fervour against imperialism as fear of the proletariat.

In the meetings of the Association in the mid-1930s one finds enthusiastic references to the views being put forward by a publication called *Isla,* not only because 'it defends our aspirations with the intention of satisfying the desires of our class' but also because 'it puts forward our views as being necessary to the happiness of the fatherland' (21 November 1936:36—37). The editors of *Isla* were Cuban progressive intellectuals who found themselves in agreement with the Communists, though they were in favour of turning all agricultural labourers into small farmers and calling themselves partisans of a petty-bourgeois policy — 'let us not make believe otherwise, even at the risk of being mocked by the Marxists'. As they explained, the Communists had given up their early 'cataclysmic tactics' and had defeated the general strike of 1935 by ordering a return to work. (Lliteras in *Isla,* 4 July 1936). One could go along with them on the anti-imperialist front. Thus *Isla* could write that the sugar industry was organized into two sections, industrial and agricultural, the hacendados (sugar mill owners) and the colonos (cane planters), divided also by nationality. The hacendados represented the imperialist form of capitalism, they lived by monopoly capital; the colonos were 'of the most Cuban origin linked forever to the destinies of our land' *(de fuente cubanisima vinculada de por vida a los destinos de nuestra tierra).* This view was supported by a 'painful document', Leland·Jenks, *Our Cuban Colony,* and by that 'dramatic book', Guerra's *Sugar and Society*(Guerra y Sánchez, 1964).

The colonos' position was nicely summed up in a bill against 'administration cane' submitted to Congress in May 1936 by congressmen who were also prominent members of the Association of Colonos ('administration cane' was cane grown by the mills themselves). The bill reads like an abstract of Guerra's book: 'Unless measures are taken to avoid the accumulation in a few hands of both sides of the sugar industry, namely the agricultural and the industrial sides, and unless the industry is maintained and restored upon the basis it has always had, the consequence will be the displacement of the native classes from the land, and therefore the proletariat will have to increase, this being a most serious danger for the stability of the economic system on which the Cuban state and the Cuban nationality are based.' Though it would be fashionable to interpret such complaints (reminiscent of Haya de la Torre's in Peru and of Gilberto Freyre's in Brazil) as a defence of a 'traditional' mode of life endangered by 'modernization', it would be more

correct to interpret them as a reaction to the effects of the
Depression on the Cuban sugar industry.

Part of the proletariat needed for the expanding production,
especially in the eastern part of the island, had been imported
from Haiti and the West Indies. One may think that the Cuban
national identity was well enough established to absorb these
immigrants without much damage, as it had absorbed the
much more numerous Spanish immigrants; one may also think
that the preservation of national identity is not all that important. It is also true that the existence of the colonos did not in
practice diminish notably the number of labourers employed
in sugar cane growing — this would have been the case had the
colonos farmed very little land, in the region of one caballería[3]
of cane each, which could be cultivated by the labour of one
family.

The appeals to the national interest came mainly from the
large colonos who employed much hired labour and who did
themselves no manual work. The colonos were right, however,
in pointing out that their own existence made for stability, as
it would be easier to unionize labourers working in plantations
than in colonias employing, say, only 10 to 20 permanent
labourers. The colonos thus acted to some extent as a paternalistic buffer against left-wing trade unionists. One should not,
however, over-emphasize this aspect as labour relations in
Cuban agriculture were rather depersonalized. It could even be
argued that the colonos found it occasionally useful to be
flanked by powerful unions, for instance, in their conflict with
the mill owners over *el diferencial* (on which more later, p.106),
and also in negotiations in Washington on sugar in the mid-
1940s.

The Communists, one would expect, would have been
pleased to see conventional Marxist predictions come true and
would have welcomed the displacement of the colonos and the
appearance of an immense agricultural proletariat. The fact
that part of this proletariat were black foreigners was irrelevant
to them. There is then at first sight something paradoxical in
the identity of views between Communists and nationalists.
However, both Communists and nationalists believed that the
complete proletarization of the peasantry would come about
through an imperialist take-over of Cuban land. There were,
therefore, consistent reasons why also the Communists could
fight in defence of the nationalist aims of the colonos. But the
moot point in this explanation of Communist attitudes is that
the nationalist line was adopted by them when the policy of

broad alliances was taken up by the International. In January 1934, the Communist-dominated union of sugar mill and sugar cane labourers was still fighting furiously the prohibition that trade union posts be occupied by foreigners and against the legislation on nationalization of labour.[4] They inveighed against the 'deportation' of foreign labourers and they asked that the anti-imperialist hatred of the masses and the struggle against unemployment be directed towards the occupation of land by labourers, the eight-hour work-day, the banning of piece-work, etc. (Sindicato Nacional de Obreros de la Industria Azucarera, 1934). One may doubt, in 1933 as in 1959, whether the anti-imperialist hatred of the masses was stronger than their dislike of unemployment. On the other hand, one may doubt whether an ultra-leftist policy had any chance of success in 1933, the Americans being ready to support first Machado, later Céspedes, and finally Batista (Aguilar, 1972).

It is the fashion today among historians in Cuba to say that the '50 percent' law was a progressive measure. One of the politicians supporting it was Guiteras, a fervent nationalist student leader, who, as a member of the government in 1933, took over the American electricity company, and who was to die shortly afterwards while attempting to engage in armed struggle against the Batista-dominated regime. Guiteras had a great influence on Fidel Castro. He would have been to the left of the Communists after 1934. But it should be noted that he himself supported Carlos Hevia, the short-lived President, on the day Batista finally took over and installed Mendieta in January 1934. Carlos Hevia had been Minister of Agriculture in the Grau-Guiteras government, was a man of the colonos and one of the architects of the very important Law of Sugar Coordination enacted in 1937. He was made honorary president of the Association of Colonos in 1951 and intended to stand as presidential candidate for the Autentico Party in 1952.

The commercial bourgeoisie was so pleased with Batista's takeover in January 1934 and his promise of law and order that their monthly review *Cuba económica y financiera* titled its leader, 'The sun has risen over Cuba'. The colonos were equally grateful to Batista and to his adviser Ing. López Castro for their part in the Law of Sugar Coordination. This legislation which came into being under Batista's nationalist auspices, put the sugar industry permanently on a quota system which discouraged 'administration cane'. Colonos were given total security of tenure, effectively implemented, and rents

were regulated and fixed at a rather low level. Payment for
sugar cane by the mills was increased. The workers and
labourers benefited greatly from the revolution of the 1930s
since they obtained social legislation and were thereafter
allowed to unionize. But perhaps the colonos benefited most
of all; this is not surprising since *all* the different factions
supported them from 1934 on. Among these factions one
might include the American New-Dealers. The valuable report
'Problems of the New Cuba' published in 1935 by the
Foreign Policy Association made the point that the colono
system was more advantageous than the 'administration
system' from the standpoint of social stability, labour relations and community life. They also rightly believed that
colonos would produce sugar-cane more cheaply than the mills.

Factors influencing the Respective Shares of Administration Cane and Colono Cane.

The quotas held by colonos and mills varied tremendously
in size.

Size of quotas held by colonos and mills, 1948

Size of quota	No. of colonos	Percent of total cane production
up to 30,000 arrobas[1]	27,134	10
30,001 — 100,000 arrobas	8,389	14
100,001 — 500,000 arrobas	3,213	23
over 500,000 arrobas	893	43
Administration cane		10
		100

[1] 1 arroba = 25 pounds

Source: L. Marrero, Geografía de Cuba, 1950:218.

There were two types of colonos according to the type of
contract with the mills, the 'controlled' and the 'free' colonos.
The former occupied land owned or rented by the mills and
were bound to sell the cane to a specific mill. The latter
farmed their own land and could sell to whom they saw fit

within the limits set by transport facilities. Both types of colonos had production quotas, fixed for each colonia and revised periodically according to crop size. These quotas could not be exceeded. Free colonos supplied a small amount of cane, in the region of only 10 percent of all cane, from the 1920s on. The controlled colonos were far more important; their share of the crop increased steadily from the 1930s on to reach some 80 percent while 'administration cane' went down from nearly 30 percent to just under 10 percent. Cuban readers of almost any political denomination would at this point be quick to point out that 'administration cane' was actually more significant than the about 5 percent which appears under this designation in the official publications of the 1950s, because some colonos were straw men for the sugar companies. This may be so, but it does not seem to have affected the shares of administration and colono cane by more than two or three percentage points. Its significance lies mainly in the fact that this was one of the minor abuses for which redress was sought.

Almost all books which have appeared on Cuba since the revolution include an impressive list of sugar companies with the mills they owned and also with the thousands and hundreds of thousands of acres they owned or 'controlled'. Some authors, such as Gutelman (1967), do not bother to correct the impression that such an impressive list may make on the reader. They do not say that nearly all this agricultural land was cultivated by colonos who could not be evicted.

There existed the common and not disinterested belief that there was a natural tendency for the mills to grow and appropriate more and more land to be farmed under the administration system. In fact it would appear that from the time the *central* (the large sugar mill) came into being at the turn of the century administration cane never exceeded 30 percent, even in the period before the 1930s. There was, of course, the very real risk that independent colonos would lose the ownership of the land through indebtedness, and this indeed had happened to some extent when the sugar market temporarily crashed after the 'dance of the millions' of 1920. However, there was no natural trend towards the displacement of colono tenant farmers in favour of administration cane, rather the reverse. There is the simple reason that growing of sugar cane was a fairly labour intensive activity, labour costs amounting to some 70 percent of total costs when cultivation and transport were carried out with oxen, weeding and

cutting with simple manual tools, loading by hand, and at a
time when very little use of fertilizer took place. Some 40 percent of the labour input was used in the cultivation process,
and the rest in the harvest. This marked seasonality of Cuban
sugar cane farming did not preclude the existence of
considerable diseconomies of scale which arose from the
difficulties in labour management during the cultivation
process.

Sugar is a cyclical crop. After the booms caused by wars
always came crises; after the expansion of credit to owner-
occupiers growing sugar cane, came foreclosures, the effects
of which were usually reduced by moratoria. Cane planters
were in the 1940s and 1950s no longer dependent on the mills
for credit. Most of them had already lost — or never had —
the ownership of the land, but they could not be evicted even
in case of indebtedness.

A factor which might have acted in favour of administration
cane was a mill's need to rely on an assured supply of cane. A
mill produces sugar more cheaply when working at full
capacity; administration cane, or cane from only a few large
colonos, made supply and its timing more dependable. On
the other hand, sugar cane being a most valuable crop, there
was really no problem in obtaining an adequate supply from
the small colonos — the problem was rather the opposite,
how to stop them from growing too much cane.

Finally, the protection given to colonos by the legislation
with regard to security of tenure, level of rents, and payment
for cane, was in itself a potential factor in favour of
administration cane. Mills were unable to displace colonos,
but they would to some extent have liked to do without
them, particularly after the 1937 legislation came into force.

The security of tenure and the regulation of rents which
the 'controlled' colonos had effectively obtained in the 1930s
was shortly afterwards extended to other types of tenants.
Through the quota system the sugar mills were under heavy
pressure, political and economic, to obtain the cane they
needed from colono tenant farmers and not by direct administration. Generally speaking, all other Cuban landowners and
tenant farmers, including the large colonos, except those
growing tobacco, a labour intensive crop, did forgo the advantages of letting or subletting land but used labourers
instead, because it became very difficult to evict tenants and
rents were controlled. The situation became increasingly
paradoxical as the number of unemployed labourers in-

creased. Landowners and large tenant farmers would have been able to profit from this unused labour by turning labourers into tenants or sub-tenants. The legislation, which was the result of militant colono pressure, prevented this trend from materializing. An observer commented on 'the latifundists' resistance to let land to cash-renters or to sharecroppers for fear of no longer being able to evict them' (Monzón, 1958).

The point was to circumvent the high wage level which did not allow landowners and large tenant farmers to profit from the work of all the labour available. In other words, the small peasant would work harder than the labourer for the same return, because labour input would not be measured against the wage level but in terms of opportunity costs. However, landowners and large tenant farmers would not profit from this increased work (in terms of days and hours of work, and in terms of effort and quality of work) if rents are fixed at a level lower than they would reach in a competitive labour market with no land tenure legislation. Cuban agricultural activities continued to be carried out, for the most part, by agricultural proletarians whose number increased due to demographic expansion, and who by and large were not given the chance to become tenants or sub-tenants owing to the legislation protecting tenants and sub-tenants. This agricultural proletariat was going to prove as dangerous for social stability as had been foreseen. Thus, although many of the colonos' aspirations were met from the 1930s onwards, in the end their policy proved to be self-defeating, perhaps a deserved result of their distasteful appeals to the national interest.

Conflicts between Cane Planters and Mill Owners

Perhaps more as a counsel of perfection than as a real possibility, the colonos came to think that they ought themselves to run the mills. The trend towards increased Cuban ownership of the mills — over 60 percent of capacity in the late 1950s as against only 30 percent in the 1920s — did not deter the colonos from thinking about the mills. Even the head of the National Bank under Batista, whom they had gone to see to discuss the financing of a project to buy the land they rented, told them: 'I do not only believe the colonos should own the land they farm; they should also own the

mills' (21 August 1954: 23). There is a continuity with the
situation which arose in 1959, when the Association of
Colonos asked for government 'intervention' at 43 sugar mills
(over a fourth of all sugar mills) on the grounds that they
were delaying payment for the cane crop.[5]

The colonos also thought they should really own the land
they farmed. It is worth considering the telling arguments
they used on one occasion (21 May 1941: 25-26). They said
that they were 'agricultural entrepreneurs', and from a
business point of view they were fairly satisfied with the
results achieved through the Law of Sugar Coordination. But
they were not only agricultural entrepreneurs, they were
also land-loving people, 'anxious to become a true landed
class'. The attack on absentee landlords was not carried out
by citing Ricardo or Henry George, but Oswald Spengler who
saw the rural exodus as a sign of the decadence of civilization.
Cuba was fortunate in having the colonos who so loved the
land that they wished to become proper landowners, not by
getting the land as a gift but rather through means respectful
of the rights of ownership. This was to be the official policy
of the Association of Colonos in 1959, when they asked the
government that colonos with less than two caballerías who,
under the Land Reform Law, would get the land free, be
allowed instead to pay for it. The small colonos themselves
failed conspicuously to support this policy.

Perhaps the main point of friction between cane planters
and mill owners was over the payment for cane, the *arrobaje*.
The Law of Sugar Coordination established that for each
hundred arrobas of cane delivered to the mill which would
yield some 12 arrobas of sugar, the colono was to get payment
equivalent at least to the value of 5.5 arrobas of sugar at the
current official price. On average, in the late 1940s and early
1950s, colonos were getting 48 percent of the value of
milled sugar. Proposals were often put forward to amend the
law and raise the share — for instance, by Fidel Castro in the
Moncada programme (see p.113). A similar conflict arose on
the payment of molasses, a by-product of sugar. The law
made no provision for payment of molasses. This grievance
became important when the production and value of
molasses grew during World War II. The colonos were
successful, after many complaints, in getting a large share in
the value of molasses produced by the mills (13 February
1945: p.12; see also Cuban Economic Research Project, 1965:
518).

A deduction of 5 percent on their *arrobaje* was made by
the mills to the colonos who rented land from the mills. This
was the moderate rent paid for the areas used for cane
farming in each colonia. Rents for the areas not used for cane
farming were freely negotiated subject only to the general
land tenure legislation. Hence the frequent demands from the
Association of Colonos for a further decrease in rent levels
for the latter areas. There were no such demands for a reduction of the rents paid for the cane farming areas, although
they are, of course, implicit in the colonos' desire to own the
land they farmed.

One further point of friction, especially in the 1950s,
arose from the suppression of *chuchos,* i.e. the transfer points
of the cane from ox-carts to railways. It was in the interest
of the mills, which paid only for railway transport, to let the
colonos incur the cost of transport to the mills or to *chuchos*
located at a reduced number of railway lines. In some cases,
compromise solutions were reached, the colonos and the mills
sharing in the costs of tractor or lorry transport which was
becoming an economic possibility. In other cases, such
solutions had not yet been arrived at in 1959. This was perhaps the most acute of the conflicts between colonos and
hacendados in the 1950s. The colonos were under pressure
from the mills to mechanize cane transport but, on the other
side, the unions were fighting against it. Another minor point
of friction was over the date when the harvest should start,
which was set by government decree. The colonos felt it was
in their interest to have an early start, in January, probably in
order to spread their labour needs over a longer period, while
the hacendados wished to start later, probably to have the
mills operating at full capacity. Also, the colonos were paid
according to cane weight but sucrose content increases with
time reaching its peak in April. Significantly, in the period
under consideration there were no complaints against the
mills on the question of credit and interest rates. There were
various sources of credit and the colonos did no longer depend
exclusively on the mills. In the early 1940s, the Association
of Colonos achieved the remarkable feat of setting up a Banco
de los Colonos which provided a further source of credit.
They also set up their own insurance company, *La Cañera.*

Finally, a grievance which colonos shared with the labourers
was that over extra payment for cane in cases where the final
market price of sugar proved to be higher than the initial
official price under which the harvest had taken place. This

H

extra payment was known as *el diferencial.* The colonos and
the labourers also got satisfaction on this point, but not without a hard struggle and occasional threats of a joint strike by
both colonos and labourers (see Federación Nacional de
Trabajadores Azucareros, 1946, for an account of one of
such conflicts).

Conflicts between Large and Small Cane Planters: Restrictionists versus Expansionists

The conflicts within the colono class itself are in close relation to the conflict about whether to expand or restrict
sugar production. The Association of Colonos was often
divided on this question but as a rule the restrictionists won.
The small colonos were under-represented in the Association.
On the one hand, the large colonos had more time to spend
on association business — the assembly of representatives
met four times a year, each time in a different province, for
meetings lasting three to four days — they were also better
educated and had the political contacts and political power
which made it advisable to elect them. On the other hand,
the electoral system worked against the small colonos by
granting equal representation to all local delegations — one
for each mill — instead of making it proportionate to the
number of associates. If the electoral system was always a
bone of contention, this was so because there was a built-in
conflict of interests between the large and the small colonos
in an industry which operated on the quota system. Large
colonos were generally in favour of restriction, small
colonos of expansion. In a laissez-faire situation small
colonos might have displaced the large colonos. Also, some
members of the small colonos' families could well be landless labourers or peasants who were not allowed to grow
sugar cane, all of whom had an interest in a *zafra libre* (a
free harvest without quotas). A bill had been submitted to
Congress in 1936 which would have made it possible for
small colonos to cut all the cane they had as long as they did
not exceed 200,000 arrobas. The Association of Colonos
strongly and successfully opposed it on the grounds that
such an important modification to the quota system then in
force — a provisional system previous to the Law of Sugar
Coordination — would cause great disruption and that the
Association as a body representing all colonos could not
favour one group over another.

The quota system established by the Law of Sugar Co-ordination was based on the quota for sugar imports granted by the United States in exchange for tariff concessions for its exports to the Cuban market. As remarked earlier, this was known as the Reciprocity Treaty of 1934. On the basis of the import quota to the U.S. and on an estimate of exports to the world market and of domestic consumption, the Cuban government — or rather, the Instituto Cubano de Estabilización del Azucar — determined the required volume of the crop and apportioned it to different mills and to the different colonias, which had as of right a *factor de molienda* (a milling factor) in the respective mill. The basis of the system was the U.S. quota, the size and price of which were regularly negotiated. Although the large colonos were on the whole in favour of this system, this did not mean that they were always happy with the trade arrangements with the United States which formed the cornerstone of the quota system. It is true that one often finds the Association of Colonos decidedly in favour of the reciprocity principle. Thus, in 1942, there was a discussion of proposals for cutting extra cane for molasses — over that being exported — to manufacture alcohol which was to be substituted for imported petrol. The Association took the initial view that 'we should not close the doors to others so that we do not find closed doors ourselves' (21 November 1942:20). Somewhat earlier they had submitted a resolution to an Inter-American Conference of Agriculture held in Mexico against artificial sweeteners which had been approved. How could Cuba now substitute imports? This is assuredly an example of the defeatist attitude one has been led to expect from an exporting bourgeoisie.

However, the colonos had at their disposal land which they did not fully utilize for cane because of restrictions on production. Thus, they quite generally took the view that substituting agricultural imports was a good policy, although they regularly claimed that more might be done along these lines if labour had not been so expensive because of union activity and generous social legislation. This support for import substitution did begin as early as the 1930s. The Association of Colonos, for instance, congratulated the Association of Stockfarmers on their attempts to make it difficult to import meat (21 November 1936: 36). Of course the congratulators and the congratulated were to some extent the same people. At least two of the six members of the

Association of Stockfarmers were also members of the
executive committee of the Association of Colonos, including
its president. Perhaps one would find a similar identity of
interests with the rice growers in the 1950s, and there is no
lack of statements of support in favour of Cuban industri-
alists. The Reciprocity Treaty was once described as a
measure by which *(las) industrias nacionales (son) relegadas
al desastre* (21 November 1944: 72).

As far as increasing exports of sugar is concerned, we find
the large colonos not too eager to expand despite their unuti-
lized capacity. They readily recognized that restriction was un-
popular with the 'misguided' working class. They found it
convenient, however, to believe that it was in the interest of
the colonato as a whole to have restriction and therefore quotas
preventing entry into sugar cane growing. Competition between colonos, they claimed, would work to the advantage of
the mills. There is evidence showing that the large colonos felt
a need to clear themselves of suspicion on this point:

> It is often said by many (small) colonos that the colonos
> who have more cane have an interest in the crop remaining
> controlled . . . They do not realize that in an open struggle
> Cuba would become an immense cane field and that the
> small colonato would be the first to disappear. We, who
> think that the crop should remain controlled, are admittedly
> a bit egotistical; but if a *zafra libre* were to take place,
> immediately the small producers would be displaced. It is
> because of this egotistical principle that we say: let us go
> on with this ordered control which protects everybody's
> rights; nobody with a right to have cane milled should be
> dispossessed of it. (21 August 1941:17).

The 'absurd idea of having a *zafra libre* without any kind of
limits' was unanimously voted down. It was not an unwise decision from the point of view of the large colonos. Despite
their restrictionist efforts some peasants took to growing cane
during World War II and also during the Korean War. The largest harvest ever before 1970 was the 7 million ton harvest of
1952 when, exceptionally, quotas did not apply. Their rights
to go on with cane cultivation were recognized by the Batista
government. Quotas had to be scaled down. The number of
colonos increased from 30,000 to 60,000 from the 1930s to
the 1950s, partly because of inheritance but also because
there were infiltrations into the colonato.

This restrictionist attitude did not exclude dissatisfaction

with the niggardliness of the United States as regards the size
of the import quota and especially price, nor did it exclude
approval for the attempts to increase exports to non-U.S. markets. However, to sum up, it would not be mistaken to say
that the large colonos constituted an exporting bourgeoisie
which was not really adverse to import substitution, but which,
on balance believed that sugar exports should not be increased
inordinately.

Nationalism and Class Consciousness

The colonos were never overjoyed by the dependent position
in which the sugar industry found itself vis-a-vis the United
States. The events of the World War II years made a lasting impression on them — an impression, one might say, of having
been taken for a ride. Towards the end of 1944 they began to
cry over the spilt milk of not having obtained as high a price
as would have been possible and of not having benefited from
the molasses export bonanza. Instead of siding with the workers
against the mill owners in demanding a share in the profits
from molasses and instead of relying on the small colonos and
on the workers to get support for selling more sugar and at a
better price to the Americans, the Association took a conservative line in the first years of the war. As late as 1944 they
refused a demand for trade union representatives to join them
and the mill owners in the negotiations in Washington — this
was only to happen later — on the grounds that 'workers are
one thing and we colonos are quite a different thing. They are
workers. We are employers, men of enterprise and responsibility, businessmen' (21 May 1944:21). The small colonos had
been duly warned that, though market perspectives were good
for the two following years, 'they should maintain their production indices, and not endanger their future in their eager-
ness for gain. We colonos are responsible citizens who, with
great effort, create the national wealth, with that sense of
building durable things which differentiates a good family man
from an adventurer, from a speculator, and from those who
profiteer from confusion'. Nothing was further from their
thoughts than to 'exhort to new investments in sugar cane
growing, or to change the "status quo" of our sugar industry
through individual or collective action' (21 February 1944:
26—27).

Aurelio Alvarez, a senator for the Auténtico Party, complained on the same occasion that the colonos were walking

hand in hand with the mill owners. Instead of putting all
possible pressure on the hacendados to get a share in the increased value of molasses, the colonos took the line that it
was impossible to grant wage increases because the colonos'
revenues had not increased. This was a mistaken strategy.
Aurelio Alvarez also discharged himself of some of his customary bitter ironies on the 'Good Neighbour': 'If they do not
want to pay the price we are asking for, and if we have not
enough dignity to stop cutting the cane and producing sugar,
then we shall have to accept the "Good Neighbour's" price
which they will be able to impose on us. But let not the Association of Colonos approach the (Cuban) government, so that
the government can then hide behind the hacendados and
colonos. As things stand now, the price they pay we have accepted; they had no need to impose it on us' (21 February
1944:33). Aurelio Alvarez himself had a quota of 1 million
arrobas. Unfortunately we cannot know what he would have
thought of Fidel Castro's revolution since he died in the late
1940s. A colleague of his, Dr. Tomás Puyans (who in 1959
appears in the executive committee of the Association of
Stockfarmers) showed himself anxious about 'the very survival
of the sugar industry menaced so mercilessly by those who
speak so loudly of a Better World' (21 May 1944:27). By the
end of the year, as shown in the meeting of 21 November
1944 and following days, the colonos as a whole had veered
round to the radical strategy: 'The colonato cannot ask for a
participation in the value of molasses unless we take it upon
ourselves to share it with the Cuban agricultural workers.'
While the colonos undoubtedly remained *el elemento neta-
mente cubano de nuestra principal industria,* the nationalist
accolade could be extended to the labourers when allies were
needed against the mill owners. If the proceeds from the sale
of molasses, suddenly multiplied in the course of the war, were
not transferred in part to colonos and labourers, the national
economy would not benefit at all. The 152 sugar mills belonged
to only 97 persons or companies, mostly foreign. What a difference it would have made if the mass of the colonato and of the
labouring classes could have shared in the higher revenue.
Their expenditure would have vigorously irrigated the national
economy.
     On the question of relations with the Americans who were
willing in 1945 to buy as much sugar as Cuba could provide
and, in stark contrast with the usual procedure, proposed not
an import quota but a 'ceiling price', the following statement
is illuminating:

> May it always be possible to harmonize the relations be-
> tween both nations, because Cuba will never be able to ignore
> the economic potential, nor her satellite status, of such an
> economic star whose influence she will feel always and for
> all time, unless an extraordinarily powerful civic strength
> liberates this people from having to appeal for affection and
> from having to resort to simulation, so that it may live the
> life of a truly free people, aware of its rights, calm and
> modest, but virtuous and positively honest'. (21 November
> 1944:12—13)

The expression 'appeal for affection' referred to the arguments
advanced by the Cuban negotiators in Washington that Cuba's
contribution to the cause of democracy had been in producing
sugar, and that the military effort had been negligible only because not more had been asked of her. The allusion to 'simula-
tion' referred to the more or less genuine threats by the Cubans
to finance storage of sugar through their own efforts and then
to wait until the Americans were forced to come to Havana
to buy. This threat, made somewhat plausible by the successful experiment in 1941 of holding on to 400,000 tons harvested
over immediate sale requirements, served as a negotiating card.

It would be something of an over-statement to say that there
is a strict continuity between complaints from the large colonos
about Cuba's satellite status and her degrading attempts to
break out of it through deference and deception and Fidel
Castro's joyful cry in July 1960 of *'sin cuota pero sin amo'*
(without quota but without a master). A socialist revolution
implied a nationalist revolution, but not vice versa.

How parochially nationalist the colonos could become is
perhaps best shown in the discussion in 1941 on selling sugar
to a Britain cut off from her colonies but with open sea com-
munications with Cuba through the Icelandic convoy route.
While Britain, or rather 'the single buying firm' in Britain, had
refused during the crisis of the 1930s to buy at even 0.70 cents
per pound, it was now offering 1.85 cents. Britain had not given
any trade advantages to which Cuba felt now morally obliged
to reciprocate.

> Cuba can follow either of two policies. The first, to make
> more and more concessions, damaging ourselves, with the
> hope that once the war is over . . . we will be better treated;
> the second one, to gain all the benefits we can from the pre-
> sent circumstances, with all justification and without extor-
> tion. To our misfortune, Cuba has usually followed the

first policy, after all to no avail, and, in my opinion, she
should adopt the second one which, after all, is an Anglo-
Saxon type of policy.

While one representative argued along these lines, another blew
his top in the following outburst: 'Cuba should not go on
sacrificing herself while there are governments [i.e. the British
government] which are spending million upon million in des-
truction for no useful purpose' (21 August 1941:62–69). At
least until the Communist Party began again to denounce imperialism after the start of the Cold War, there was no nationalism in Cuba comparable to such vehement bourgeois nationalism.

There is a view of Cuban society, propounded by Robin
Blackburn in 1963 and taken up by a number of writers, among
them Hugh Thomas (1967 and 1971), many of them basing
their opinion on the extremely weak basis of Lowry Nelson's
views on Cuban class structure (1950), i.e. that there was no
nationalist middle class of any consequence. Hugh Thomas
believes that there existed an 'intellectual', 'professional'
middle class which attempted to pursue freedom from sugar
and from the United States, but he denies the existence of a
nationalist business middle class (1971:1111–2, 1236–7).
Another writer asserts that 'the only national class in Cuba
was the working class' (1969:61). Robin Blackburn asserts
that there was 'a proto-bourgeoisie (which) never achieved true
consciousness of its interests and identity; lacking elementary
class consciousness, it never discovered class solidarity. It produced no institutions. Lumpenized, destructured, disintegrated,
it failed to lodge itself lastingly in Cuban history. When the
guerilla troops entered Havana on January 1, 1959, its extinc-
tion was close' (1963:73–74).[6] This is a view which is not
quite right. Some strong institutions were produced, among
them the Association of Colonos. Material goods the colonos
might have lacked, but they were oversupplied with class consciousness and frustrated aspirations which they identified with
the national interest. They hoped, or at least half-hoped, that
the revolution of 1959 would fulfil these aspirations. The
period after 1952, when drastic restriction on sugar output
was reintroduced since the Americans would not buy nor pay
more for it, disappointed them.

The colonos saw themselves as nationalist, middle-class businessmen. On occasion they went so far as to recommend in
their summer meetings the use of *guayabera*, a Cuban shirt to
be worn without a coat or tie. As they once said, 'the Cuban
people will have to go on seeing in the Association of Colonos

the spearhead of a leading national movement, *los empresarios de la tierra criolla'* ('the businessmen of this Creole fatherland') (12 May 1945:147). It is then a remarkable mistake to say that the 'nebulous stratum between the upper and the lower echelons of Cuban society was made up not of one but of several sectors, none of which, either separately or as a whole, had a consciousness of class' (Ruiz, 1968:13).

The Revolution — the Moncada Programme of 1953 and the Land Reform Law of May 1959

A cursory analysis of the Moncada programme shows the keen interest on the part of Fidel Castro for the problems of the colonos. This is not surprising since the colonos occupied one of the main places of Cuban political life. There have been writers on Cuba, baffled by details, who have interpreted Fidel Castro's Moncada proposal to raise the arrobaje to 55 percent of the sugar yield as a proposal directed to remedy the situation of some downtrodden sugar cane sharecroppers. Arrobaje was a question much discussed at the time, and a Comisión Técnica Azucarera convened some months before Batista's coup of August 1952 had produced precisely in 1953 a monumental report on sugar growing and milling costs to serve as a basis of agreement between hacendados and colonos. In a meeting in 1954 there is a disclaimer that the colonos were aiming quite so high: 'when we discussed an increased share in the yield of our canes, the executive committee and the assembly knew how far we could go, though the figure of 55 percent was mentioned' (21 August 1954:17). But it was fully consistent with a pro-colono standpoint and well within the realm of political reality to propose an arrobaje of 55 percent. This had been first proposed in 1948 by the Communist Party, since 'the colonato's struggle should be included as a whole in the objectives of the struggle for National Liberation whose main enemies are the foreign imperialists and the semi-feudal latifundists' (Pino, 1948).

Another of the points of the Moncada programme was devoted to the small colonos, i.e. those with quotas under 30,000 arrobas, who were to be allowed to sell all the cane they had up to 40,000 arrobas, provided they had been established as colonos for at least three years, thus excluding the recent large intake of the 1952 zafra libre. This was a proposal within the main-stream of Cuban radical political thought, but not necessarily alarming for the large colonos since any permanent extra quotas for the small colonos could have come from what still remained of administration cane and from any future expansion of the crop. This Moncada proposal was put into effect in 1959 with strong support from the Association. More alarming was the line taken by Guevara, as reported by the executive committee of the Association; he considered that any permanent increases in the small colonos' quotas should come from the large colonos, since the quotas hitherto used by administration cane should now go entirely to the new cooperatives. And, while in 1959—60 the Association was in favour of allowing the small colonos to sell all their cane up to the 40,000 arrobas limit, this being a 'just purpose which has always inspired the actions of the assembly of representatives in defence and help of our most needy fellow-members' (February 1960:5), similar measures had not been supported in the past. A proposal in 1941 to allow the small colonos to cut up to 50,000 arrobas, taking the extra from administration cane and from those colonos whose cane exceeded 10 percent of the total cane milled in the respective mill, was not adopted by the assembly on the grounds that it would be disliked by the hacendados with whom a joint negotiating trip to Washington had been planned for the near future.

The Moncada programme also proposed land reform by promising to implement the constitutional provision of 1940 which proscribed latifundia. In principle in Cuba the sugar companies' immense properties — most of them actually farmed by smaller holders — and the big cattle ranches of Camagüey, were classed as latifundia. The land reform law of May 1959 appeared to fulfil the long standing desire of the colonos to get the ownership of the land they farmed. It also laid down that no person or company could own more than 30 caballerías, a limit which could be extended to 100 in some cases (1,000 and 3,300 acres). As implemented, arable holdings were usually given a 30 caballerías limit and pasture holdings one of 50 caballerías. The colonos were thus favoured and even the largest of them, unless they owned also pasture land in large quantities, had little to lose from the 1959 Land Reform Law. Had the law been applied literally, at most it would have split the large colonos — approximately 1,000 who grew more than about 10 caballerías of sugar cane on holdings of more than about 20 caballerías and with quotas of over 400,000 arrobas — into four-fifths in favour and only one-fifth against the law.

It is a sign of the government's pro-colono mood in 1959 that the land reform law explicitly stated in its last article,

No. 67, perhaps included as a generous afterthought, that colonos with over 5 caballerías (165 acres) would be entitled as of right to buy the land they farmed up to the 30 caballerías limit.

It was not until 15 January 1960 that a Permanent Commission to apply the land reform law to the sugar cane areas came into being. The colonos and the hacendados were informed that a local commission would sit in each sugar mill constituted by a delegate from INRA (the Land Reform Institute), a delegate from the Rebel Army, the president of the local delegation of colonos affiliated to the national Association, the secretary of the local branch of the union of sugar mill workers and agricultural labourers and the sugar mill administrator. Negotiations took place between the colonos and the Permanent Commission in Havana on a number of points on which the meaning of the law was open to doubt, namely:

1 what price would be paid for the expropriated land, and also for the standing cane?

2 would the colonos exceeding the 30 caballerías limit be allowed to choose the location of the 30 caballerías they could keep?

3 how was the cane quota to be distributed between the expropriated and the 'allowed' area?

4 under what terms would the colonos be able to buy the land they farmed up to the 30 caballerías limit?

On point 1 the colonos were ready to accept the very much undervalued tax assessment as basis of payment for the expropriated land, but they expected to pay for the land they would buy on the same basis (point 4). They hoped to make some money not so much through compensation for the land expropriated but for the standing cane plantations. They asked for $ 6.00 for each 100 arrobas of cane, i.e. something in the region of $ 2,500 per caballería of sugar cane (one Cuban peso = U.S. $ 1.00). The government commissioners found this slightly expensive and proposed to pay on the basis of the 1953 Comisión Técnica Azucarera studies, which the colonos considered out of date as regards plantation costs.

On point 2 the colonos demanded the right to choose the location of the 30 caballerías they could keep as landowners or acquire as renters. This was, of course, not a specific colono problem. All other landholders exceeding the limit were con-

fronted with the same question, which was decided by the land
reform authorities taking into account such considerations as
the landholder's political record, local pressure on land, etc.
Quite often landowners and renters were given the right to
choose the location of their 30 caballerías (or 50 for pasture
land). They naturally chose the central buildings of the estate,
the best soils and, in the colonos' case, the valuable sugar cane
area. The local sugar mill commissions seem to have worked on
the basis that the sugar cane area should be apportioned pro-
portionately between the 30 caballerías 'allowed' area and the
excess area, but there are examples of colonos being allowed
to keep up to the full 30 caballerías in sugar cane fields. At
harvest time, this would require the employment of some 50
cane cutters. The fact that in January 1960 the colonos still
demanded the right to choose, shows that they felt this was
still an open question, and this view seems to have been shared
also by the government commissioners.

On point 3, closely related to point 2, the colonos asked
that Article 59 of the land reform law be applied flexibly. This
article stated that the quota went with the land, which was
ambiguous when it was a single colonia which had to be divided
up. The commissioners seem to have agreed that at least a part
of the quota equivalent to the share of the 'allowed' area in the
total area be kept by colonos whose holdings exceeded 30
caballerías limit. However, in cases where colonos were able
to keep the sugar cane area and hand over the rest, they were
presumably allowed to keep the entire quota.

On point 4 the government commissioners explicitly agreed
that colonos should be able to buy the land for exactly the
same price which was being paid by the Land Reform Institute
to the owners. They also agreed to colonos spacing payments
over a 20-year period. When the land farmed by the colonos
belonged to landowners owning less than 30 caballerías — a
rare eventuality since most colonos farmed land belonging to
the mills — the land was also to be expropriated by INRA
(according to Article 6 of the law) and sold to the colonos. In
this case the colonos could enter directly into negotiations
with the owners to buy this land in order to save time. If agreement could not be reached, INRA would intervene in favour
of the colonos. Although the colonos paid low rents — less
than 100 pesos per caballería per year, for the sugar cane
area — buying the land on such favourable terms was something
to be desired.

The only issue about which the colonos had reason for ser-

ious misgivings after the January 1960 negotiations was over
the enactment of the necessary regulations before they could
buy the land. The timing for these regulations, which in fact
were never enacted, had apparently not been discussed. Colonos
farming between 5 and 30 caballerías of land were not allowed,
after all, to buy the land they farmed and expropriated by INRA
from the sugar mill companies despite Article 67 and the agreement reached on point 4. Holdings over 5 caballerías not already taken by the government were eventually expropriated
in October 1963. The colonos farming them had already had a
rough deal in previous years. However, harrassment of these
farmers does not appear to have been the initial policy of the
revolutionary leaders. Thus, in the forum on land reform in
July 1959, convened by the 26th July Movement (Fidel Castro's
political group), farmers holding between 5 and 30 caballerías
were referred to as *campesinos* (peasants) by Marcelo Fernández,
then national coordinator of the 26th July Movement and
Minister of Foreign Trade in the late 1960s. Perhaps a more
useful definition of that social group was given by the Communist Party spokesman on the same occasion, when he explained that the enemies of the revolution were the imperialists,
the latifundists, and the large importers; other classes were all
interested in the independent national development of Cuba,
and were to march united in the present stage of the revolution: the workers, the peasants, the middle classes, and the
national bourgeoisie, both in its industrial and rural varieties
(see Primer Forum Nacional de Reforma Agraria, 1960:528,
560).

Infighting in the Association in 1959

It was in keeping with the pro-colono character of the land
reform law that it forbade farming by corporations unless all
shares belonged to Cubans and shareholders were not at the
time shareholders of a sugar mill company. This was a further
victory in the war against administration cane. Although the
Moncada programme and the legislation which arose from it
in 1959 could be partly described as a collection of pro-colono
measures which scarcely discriminated against the large colonos,
it would be mistaken to think of the Association of Colonos
as a bulwark of the revolution even in 1959. The Association
of Colonos, as a pressure group, had learnt to live with all re-
gimes, and some of the members of its executive committee
and of the assembly were men picked to deal with the Batista

regime. The executive committee had congratulated Batista on the failure of the attempt on his life in March 1957, as almost all other producers' associations and civic bodies felt obliged to do. In the noisy meeting on 16 March 1959 it was announced that Comandante Fauré Chomón, one of the few survivors of the assassination attempt, would be 'pleased' to come on the next day and meet the members of the Association who had gone and seen Batista. There was a proposal to disqualify for 30 years members of the assembly who had voted in favour of attending the ceremony in Batista's honour. The proposal, put forward by a Dr. Ruiz Leiro, would have meant a drastic change in personnel, but it was never brought to a vote mainly through the efforts of Dr. Sardiñas who argued that to disqualify representatives for political reasons would infringe the Association statutes. As a lawyer he was against this, and he knew that the leader of the revolution, himself a lawyer, would also be opposed to it.

The defection of some members of the provincial and national assemblies of the Association of Colonos who fled with Batista had made new elections imperative early in 1959. The elections were supervised by Rebel Army members and many old regime colonos to be found in the meeting of March 1959 were then newly elected. It was said later that although the elections had been democratically conducted, the election statutes themselves were undemocratic, discriminating against the small colonos, and this old grievance was later to be the excuse for finally disbanding the Association of Colonos in 1961, when the Association of Small Agriculturalists (ANAP) was substituted for it. For the time being, however, the colonos were able to withstand pressure. In the 15-day-long forum on land reform of July 1959 the Association was represented by Dr. Sardiñas who duly supported the land reform. Dr. Ruiz Leiro also attended as self-invited and self-appointed head of a so-called Free Association but not much attention was paid to him, and he was allowed to speak only once, at 3 a.m., after the session had been formally closed. Dr. Sardiñas was to be the most active public spokesman for the colonos throughout 1959. Towards the end of the year he represented the Association in an appearance on television with Conrado Becquer, the head of the union of sugar mill workers and agricultural labourers, when he once again gave *una demostración de amor a nuestra clase* (November 1959:18).

The meeting on 16 March 1959 was adjourned for a few days to let the excitement calm down, and a new executive committee was then elected. Members of this new executive committee who had been long active in the Association and who were to some extent compromised with the Batista regime were Dr. Ramiro Areces, the secretary, and Sr. José Pérez San Juan. In this happy ending to their political differences, the colonos were helped by the 'formidable diplomatic efforts' of Ramón Castro Ruz, who was 'not only a revolutionary, but also had the no less honourable quality of being a colono' (minutes of meeting begun on March 16, 1959:17). The colonos thought they had found among their number, as they were used to finding, a man suitable to the new political situation and a loyal member of the colonato.

Sugar Policy

Commentators on the Cuban revolution have usually interpreted Fidel Castro's pronouncements in favour of diversification of the economy as though they implied that much less emphasis was to be put on sugar production. This hasty conclusion gains force from the fact that this is what actually happened, culminating in the very poor harvest of 1963. A reduction of sugar output, or even maintaining it at the late 1950s level, might have meant, in the new political situation of 1959, a reallocation of cane production in favour of small colonos and in favour of the new cooperatives set up on some of the expropriated lands at the expense of the medium-sized and large colonos. Thus, it would be inconsistent to state that Fidel Castro had a pro-colono policy and that, at the same time, he was planning a reduction in sugar production. This interpretation of Cuban history sees the 1959 revolution as a step in a long line of attempts at pursuing freedom from sugar and from the United States. The discussion has to be set in the context of the evolution of the Cuban economy from the 1930s to the 1950s, of which no more than a very brief account is needed here.

In that period Cuba emerged from the Depression and her GNP doubled from the level of the 1920s. Population was growing at about 2 percent per year reaching 7 million in 1959. There was no immigration, but the beginning of emigration to the U.S. in the 1950s. Sugar production did not expand above the level reached in the 1920s. Sugar milling capacity was not increased, but was in fact under-utilized. Cuban mills were able to produce 7 million tons per harvest season but their output rarely exceeded 5 million. Per capita income did not grow above the level of the 1920s. The growth of the economy was achieved through diversification of two kinds: production for export, especially of minerals (nickel) and expansion of consumer goods industries (textiles, shoes), and of food production. To a considerable extent revenues made during World War II and the Korean War were used to acquire foreign assets in Cuba and for repayment of debts to American banks incurred during the crisis. It is probably the case that between 1940 and 1958 the Cubans nationalized American investments in Cuba of greater value than after 1959, such nationalizations being carried out by private citizens and companies. Though there were reasons for a feeling of success, there was widespread dissatisfaction over the stagnation of per capita incomes, the increase in unemployment, the failure to increase sugar production, etc.

Going back now to the question of sugar policy in 1959, Hugh Thomas believes that the new administration had no expansionist sugar policy in 1959. There are a few important points which militate against Hugh Thomas' interpretation. For instance, the inclusion of Cepero Bonilla in the Cabinet in January 1959 only makes sense if sugar expansion was planned, and it is not surprising that Hugh Thomas does not mention Cepero's *Politica azucarera* (1958) in which the case for expansion was vigorously argued. Further, the statement by the production director of INRA, Oscar Pino Santos, in the summer of 1959 that Cuba would follow an 'aggressive sugar policy', is dismissed as 'meaningless' (Thomas, 1971:1240) while Fidel Castro's proposal to the United States in June 1959 that they should increase substantially the Cuban sugar quota, up to 8 million tons, is discounted as 'not . . . practical' (ibid.:1224). This proposal was obviously over-optimistic but it is at least a sign that a policy of increasing sugar production was being explored. In May 1959 sugar prices came down partly because of 'Castro's airy talk of producing as much sugar as possible and of selling it below prevailing world market prices' (ibid.:1214). Therefore, it is misleading to conclude that the 'Cuban radicals' inclination in 1959 was not to spend too much time on sugar, rather on diversification of agriculture and on industrialization. The great aim was to escape from monoculture' (ibid.:1154). For they believed that both, expansion of sugar production and diversification, were compatible because of the existence of unused resources. Fidel Castro himself, in the Moncada programme, mentioned unused savings, unemployment, and unused land: this was a scandal which transcended the economists' circle.

Thus the existence of an 'historically comprehensible . . . obsession to escape from sugar' (ibid.:1325) during the years 1959—60 before the U.S. government's threat to cut the quota, is largely in Hugh Thomas' imagination. Few believed that 'sugar had taken Cuba to an *impasse*' (ibid.:1155), but many believed that *restriction* of sugar production had, since it had meant a decreasing Cuban share of the world market. Viriato Gutiérrez, co-author of the Chadbourne restrictionist plan of 1932, had long recanted publicly. Julio Lobo's initial support for the revolution — paradoxical in that Lobo was Cuba's richest mill owner — was due to his hope of seeing implemented the expansionist policy he had long preached.

The factors which favoured long term growth in sugar production were well known and often mentioned in Cuba, such as the high income-elasticity of demand over a long range of incomes, and the fact that Cuba could, and can, produce cheaper sugar than almost any other country. Against expansion there existed, and still exist, the protectionist policies and special arrangements in the world sugar markets, and the constant fear about the development of artificial sweeteners in metropolitan countries. Discussions on restriction often revolved around such themes. The radicals' explanation of restriction was that it had served the interests of the speculators. Fidel Castro said in a speech in 1965: 'Everybody knows the story of sugar restriction. We produced 4 million, 4½ million, 5 million tons, and the cane was left over for next year and workers had no work. For the sugar barons were only interested in the ups and downs of sugar prices and in speculating in sugar' (in his speech of 7 June 1965). But there is room also for an explanation of restriction not in terms of the Cuban 'national interest' but of sectional interests. As already suggested before when discussing the conflicts between large and small colonos (p.108) a possible explanation for the permanence of restriction is the fact that both, small sugar mills mostly owned by Cubans and large cane planters, were in favour of it. If quotas were removed they could not compete with the large sugar mills and small cane planters, as there are economies and diseconomies of scale in the industrial and agricultural sides of the sugar industry. Their alliance proved more politically powerful than the alliance between large sugar mills, some of which were still American, and small cane planters. Be that as it may, by 1959 few people in Cuba dared to defend restriction.

The fall in sugar output was not at all an intended result. Brian Pollitt's remarkable work (1971) shows how the govern-

ment forgot about the implications of the seasonality of
agricultural work. The unemployed agricultural labour force
was not as readily available for alternative uses as was thought.
Also, perhaps the government were not radical enough — or,
more accurately, the broad class alliance did not permit — to
concede wage increases to labourers sufficient to keep them in
the agricultural wage sector, now that the threat of seasonal
unemployment had started to disappear and, in the confusion
of the land reform, little plots of land could, within limits, be
acquired. The end of the incentive to work hitherto provided
by seasonal unemployment meant a decrease in effort and
availability of labour at the peak of the harvest. This problem
still plagues the Cuban economy.

The labourers increased enormously in political strength in
1959 and were able to put pressure on the government to
broaden the land reform beyond its initial scope. However,
the revolution was not yet radical enough to realize that both
for social and economic reasons there would be a need to
change the valuation of agricultural jobs, especially of cane
cutting. The government maintained considerable wage differentials which were biased against agricultural work. This resulted in a shortage of manpower for the sugar cane harvest
with no comparable benefits in other sectors since labour
was often used for pursuits of dubious economic value. The
resulting fall in sugar production was unintended.

The Conflict with the Labourers in 1959—60

How persistently self-confident the cane planters still were
early in 1960 is shown in their proposals on labour relations
submitted in January in a memorandum addressed to the
Minister of Labour. Answering the demands from the
Federación Nacional de Trabajadores Azucareros for a rise in
social security payments to labourers who suffered accidents
in work, they pointed out that they were unable to contribute
any higher premiums. They were nevertheless not averse to an
increase in benefits if the state would take care of them, provided they were not set so high that they became an incentive
to self-inflicted wounds. Indeed, such cases were not unkown
but it was not a wise observation to make in public in 1960.
Social security for agricultural workers was well developed in
Cuba. It had begun in the 1930s and was expanded later. When
the revolution came it provided retirement, work accident and

maternity benefits, but not medical nor unemployment
benefits.

In the same January 1960 memorandum the colonos took
the opportunity to attack the mill owners by asking for an increased payment for cane. This was the only way by which
the colonos would have been able to satisfy to some extent
the labourers' wage demands. An increase of arrobaje by 19.37
pounds of sugar (i.e. approximately 0.8 arrobas, or an increase
of 13 per cent) for every 100 arrobas of cane delivered to the
mill would have covered, at the current official price, only 50
percent of the increase in agricultural labour costs experienced
since the enactment of the Law of Sugar Coordination back in
1937. Unless this demand was granted, the colonos would have
been hard put to pay any extra money to the labourers. The
señores hacendados were exhorted to realize that the colonos
constituted a hard working class of 60,000 members which
should get an economic reward commensurate with the importance of its social function. If this increase in arrobaje would be
granted, the colonos were ready to accept an increase in wages
for the harvest labourers of 14 million pesos, which for a harvest of 4 million tons meant 35 cents per each 100 arrobas of
cane, equivalent to nearly 10 pounds of sugar at the current
official price. This would have meant a further increase in
labour costs over a similar one the colonos claimed to have
endured in 1959.

Even though the increase in arrobaje was granted, the colonos
nevertheless refused in the case of left-over cane to pay any
extra money over the established piece-rates, this being one of
the grievances of the labourers. With regard to cane in fields
where yields per caballería were less than 30,000 arrobas, the
colonos conceded the point that it was difficult for cane cutters to reach sufficient earnings. It was decided that unless the
low yield was due to drought, negotiations over higher piece-
rates for such fields should be carried out in each specific case
between the colono and the local union. There was a perpetual
conflict over the rates for harvesting left-over cane, cane in
fields with a poor yield, and insufficiently weeded cane.
Labourers sometimes resorted in 1959—60 to burning cane
fields in order to get colonos to agree to cane being cut where
cutting would have been uneconomic; cane cutting is easier in
a burnt cane field but it must be done at once if the cane is
not to spoil.

The colonos became perfectly aware in 1959—60 that their
survival as a class depended on providing employment for the

mass of the unemployed, but wage increases, labour unrest, the breakdown in the credit system and, sometimes, the delay in payment by the mills, made this difficult for them. 'The executive committee has tried to make all colonos aware of the need to make the maximum effort to give work to the labourers in order to mitigate as much as possible the crisis that our labourers are confronting in these last stages of the off-season. We know that our budgets and also credit facilities are practically exhausted but nevertheless we reiterate to the assembly the need to exhort our associates, as we ourselves have done, to help the labourers as much as possible even at some cost to us' (November 1959:11).

A frequent form of conflict in the first months of the revolution was over wage claims presented by the labourers with support from the local union branch. Disputes such as these, and also disputes over *tiro* and *extra-tiro* (payment to carters for the transport of cane) had a tendency in 1959 and 1960 to be brought by the labourers to the Land Reform Institute offices where they were often solved by resort to 'intervention'. 'Intervention' took place with great frequency when the dispute was over unemployment. Officials from INRA were under pressure from labourers to take over farms and create employment. Hence the suggestion from the colonos in the January 1960 memorandum, that disputes be submitted exclusively to the Ministry of Labour and not the INRA offices. They tried to separate disputes on economic matters, such as wages and employment, from land reform, i.e. from the question of ownership and control of the means of production. But in the conditions of political uncertainty obtaining in Havana, these were really two sides of the same coin. Pressure from the labourers for work or land made the land reform of 1959 go beyond its initial purpose.

Conclusion

Ramiro Guerra had prefaced *Sugar and Society* in 1927 with a quotation from Pozos Dulces in 1866: 'a race which abandons the cultivation of its territory to other races deprives itself of any legitimacy of possession.' In 1866, 'other races' did not only mean African slaves, but also Spanish merchants turned mill-owners who were displacing the 'white Creole' planters. In 1927, 'other races' meant Haitian and Jamaican immigrants, but also, more importantly, American imperialist mill-owners who were taking an increasing hold over Cuban land. The struggle of the colonos against such imperialistic forces found its emotional narrators in Ramiro Guerra and Fernando Ortiz. The struggle between the colonos and the labourers went unsung but it was to prove decisive in the end. The two conflicts were not unrelated but they cannot be subsumed under the heading 'The Cuban Nation against Imperialism' nor 'Cuba or the Pursuit of Freedom'. 'Cubans have . . . been . . . in love with the word "liberty"; slaves sought liberty from masters, merchants from Spanish laws, romantics from the Spanish army, twentieth-century intellectuals from the strait-jacket of sugar' (Thomas, 1971:1491). But 'intellectuals' should not be mistaken for spokesmen for the cane planters, and especially the strait-jacket of sugar for the strait-jacket of the sugar mills.

The mark of an objective interpretation of Cuban agrarian history would be a reluctance to believe that if Cuba became a 'Plantation' it could never become a 'Nation'. In the nineteenth century a plantation society could well have become a nation of ex-slaves, an abhorrent prospect to the white Creole haunted by the Haitian experience, while in the twentieth century a plantation society could have become a nation of proletarians, a no less abhorrent prospect for the bourgeois nationalist. Hence the prevalent interpretation in terms of the dilemma 'Nation or Plantation'.

This new interpretation, instead of denying the reality of the political power of Cuban mill owners, Cuban workers and, especially, Cuban cane planters, and instead of glossing over the Law of Sugar Coordination — dismissed in one paragraph by Julio Le Riverend — is able to acknowledge the importance of this legislation as a most decisive result of the 1933 revolution. It is also able to take into account the fact that this legislation was to some extent counter-productive. This was not because sugar 'required' a plantation system, but because the legislation protecting tenants and regulating rents discouraged sub-division of holdings, thus failing to prevent the rise of a large and militant agricultural proletariat. If the socialist revolution was successful, this was perhaps not so much because of the absence or weakness of a nationalist bourgeoisie as because of the strength of the working class. The radicalism of the bourgeois nationalists tempered by their fear of the proletariat but urged on by their conflicts with the United States over sugar policy, was also a factor in the 1959 revolution. It took them a while to realize that this was not the revolution they had been hoping for — in such a heady moment, they had forgotten about the proletarian threat, which did materialize soon after January 1959.

In this paper it was not my intention to study the reasons for the radicalization of the land reform (see the next essay in the present volume and Martinez-Alier, 1972) nor to give a detailed account of the vicissitudes of sugar production either before or after 1959 (see Cuban Economic Research Project, 1965 and Pollitt, 1971). The purpose was to provide an explanation of why a socialist revolution triumphed so easily in Cuba, an essential part of which rests on the given analysis of the social position and ideology of the cane planters.

NOTES

1. A competent summary of the evolution of the sugar industry may be found in Levi Marrero (1960:201–229). Detailed information on the sugar industry is given in Cuban Economic Research Project (1965).

2. The main source is the minutes of the quarterly meetings of the assembly of representatives of the Association of Colonos, and the reports of its executive committee, cited in the text by their date. All the materials quoted are in the National Archives or in the National Library, Havana.

3. About 13.5 hectares (33 acres) producing some 40,000 arrobas of sugar cane (1 arroba = 25 pounds).

4. The notorious '50 percent' law of November 8 1933, required any employer to employ native Cubans in a proportion of at least 50 percent of his total number of employees. Similar legislation was enacted in Argentina and in Brazil in the 1930s.

5. Cuba económica y financiera, August 1959. The petition did not come from a group of small planters, as Hugh Thomas believes wrongly but consistently (Thomas, 1971:1240).

6. Cf. J. O'Connor (1970) for a better informed interpretation.

V

The Peasantry and the Cuban Revolution from
the Spring of 1959 to the end of 1960

In 1959 there were in Cuba about 500,000 agricultural labourers, 100,000 small tenants of various types, and 100,000 small
peasant owners. I shall first give a short account of the conflicts
that opposed landlords and small tenants. Then I shall proceed
to study the more important conflicts between landowners
(and farmers) and the labourers, and this will provide the
occasion for some discussion of the writings of Draper and
others. I shall also deal with the often-made assumption that
peasants want land while proletarianized labourers want higher
wages and good employment. Finally, I shall explain the reasons behind the present drive for collectivization, which is to
some extent helped and to some extent hindered by decisions
taken in 1959–60.

Had the Plantations displaced the Peasants?

Let us first consider the landlords and tenants, which in
Cuba meant to a large degree the American sugar companies
and Cuban colonos (the sugar cane growers). Outside Cuba
the impression had been given that the sugar-mill companies
were themselves in charge of the agricultural operations. It
was unfortunate that Ramon Guerra's *Sugar and Society* was
translated into English after the Cuban revolution without an
explanation to the reader about how the situation had changed
from the 1930s revolution on.[1] Guerra's book has been described as a 'good case study of the replacement of peasants by
plantations' (Wolf, 1966:12). One must keep in mind, however,
that this book, very influential in Cuban politics, deals with
the situation in the early 1920s and even then it would have

been an exaggeration, for the sparsely populated regions of
Cuba, to say that plantations were displacing peasants. In some
other regions, it is true that after the sugar market crashed
following the 'dance of the millions' of 1920 and after the
first restrictions on sugar output were introduced in 1926,
some sugar-cane growers were replaced by administration cane
owned by the factories. They lost the land they had mortgaged.

In the 1930s the colonos mustered sufficient political force
to create a powerful association and, riding the nationalist
tide, they achieved favourable terms in their dealings with the
sugar companies. Although they were unable to recover the
ownership of the land they had lost, they achieved, among
other concessions, complete security of tenure, regulated rents,
and a fair share of the value of the sugar. Not in vain had they
found Guerra's book an 'admirable and patriotic' work which
defended *la clase más cubana,* themselves,[2] and which drew
attention to the political dangers of proletarianization. Simultaneously, or not much later, all peasants — whether sugar
cane producers or not — got security of tenure and limitations
on the level of rents.[3]

But the harm was already done. *Desalojos* (evictions) became a 'hot' issue in Cuban politics. Despite legislation, some
peasants — though very few growing sugar cane — were now
and then evicted in the 1940s and 1950s, sometimes because
they had installed themselves on private land alleging that it
was public land, sometimes because they had ceased to pay
rents. Sometimes they left the land they occupied but received
a substantial indemnity — this was known as *vender la acción.*
And there were, of course, some genuine illegal evictions,
much publicized, which had the function of keeping desalojos
on the boil. In the 1959 land reform law *desalojados* (those
who had been evicted) were given priority in getting land.
From my study of the Agrarian Reform Institute papers it
would appear that many desalojados who had not been the
victims of grave injustices received land. Some had got indem-
nities which equalled a whole year's income, a few seem to
have been professional desalojados — persons who settled down
on somebody's land and refused to move unless they were
paid, repeating the operation time and time again; the alternative for the unfortunate landowners was a court case.

It is not difficult to understand why Cuban radicals — including *Bohemia,* the Communist Party, and the Moncada programme — blew up the issue of evictions out of proportion
during the 1940s and 1950s. They were giving notice to the

American sugar companies that the settlement established in
the Ley de Coordinación Azucarera of 1937 had come to stay,
and would be made still more favourable to the colonos; never
again would the Cuban nation allow plantations to displace
colonos, and all land should be the property of Cubans. They
were expressing their belief in the necessity of a numerous
Cuban peasantry and therefore registering their protest when
its ranks suffered a loss. The Communists, once they renounced
or were made to renounce dreams of revolution after 1934,
made the defence of the peasantry and of anti-imperialism the
cornerstone of their programme for alliance with the 'national
bourgeoisie'. The radicals also felt it incumbent upon them to
make a row every time an illegal desalojo took place because
it had been Batista who had promoted, either while in office
or as head of the army, much of the legislation protecting
tenants; genuine desalojos, evictions of peasants without due
process of law by the Guardia Rural ordered into action by
geófagos, provided useful political capital. Such evictions
proved that legislation meant little if the administration was
corrupt: the creed of the Ortodoxo party.

It would seem, however, that the conflict between landlords
and tenants over security of tenure had been reduced to man-
ageable proportions before 1959. Landlords and tenants (including the cane growing colonos) did continue to have conflicting
interests on the level of rents, and here the legislation had not
always been enforced to the same extent. This conflict came
into the open in 1959. The land reform law gave ownership
rights to all kinds of tenants: cash tenants, sharecroppers, etc.
But the transfer of ownership rights was to be done by the
Agrarian Reform Institute, after expropriation, and this procedure could take some months or even years. In the meantime
it was not clear in the law whether the peasants had to go on
paying rents and shares. The Agrarian Reform Institute had to
decide what instructions to give to its local officers in this
matter. There was some vacillation because peasants did not
wish to pay, landlords felt still strong enough to become indig-
nant, and the Agrarian Reform Institute was trying to comply
with a law which was itself very ambiguous. It is significant
that instructions came from the Agrarian Reform Institute
stating that payment of rents in sharecropping — and also in
emphyteutic tenure — could now stop because these were
'semifeudal' forms of land tenure. Payment from cash tenants
must continue to be paid. In practice, however, it would seem
that tenants — at least, small tenants — ceased to pay rent.

K

The Demand for Work or Land

More important than conflicts between landlords and tenants were conflicts between landowners, and large farmers, and the labourers. Their study offers the opportunity for some reflections on the interpretations of the first years of the Cuban revolution put forward by Draper and Andrés Suárez.

First Andrés Suárez. This is from an abstract of his book:

Foreign policy determined Castro's conversion to Communism. The 'peasant origin' of the Cuban revolution is no less a myth than the assumption of an active part taken by the C.P. The picture arises of a movement shaped largely by the impact of one man, whose purpose — to extend the revolution to other Latin American countries — made him join the Communist camp.[4]

To repeat: foreign policy determined Castro's conversion to Communism. However, Suárez's discussion of the agrarian question is poor, and I would suggest that a proper analysis leads one to the conclusion that the peasants played some role in Castro's conversion to Communism — not so much before January 1959 as afterwards.

While dismissing the importance of the peasantry and excluding from it the agricultural labourers, Suárez comments:

It is difficult to see how a segment that did not amount to 6 percent of the economically active population and plainly did not have the avid hunger for land that was later discovered by observers (when Guevara started the myth of the peasant revolution) could have made an important contribution to the revolutionary situation in January 1959 . . . (1967:34)

But, later, he says

It should be noted that in this underdeveloped country neither Guevara nor the Communists were demanding wage increases or making any other demand on behalf of the workers who, if one includes the agricultural sugar cane labourers, made up the great majority of the Cuban people. (ibid.:42)

The peasantry, therefore, if one includes the agricultural labourers, was not after all such a small segment of the active population; perhaps some 35 percent made up, as I have said, of approximately 100,000 small owners, 100,000 small tenants of various types, 500,000 agricultural labourers.

Two points may be made at this stage. First, as Suárez correctly says, neither the extreme leftist Guevara nor the PSP put forward demands on behalf of the agricultural workers. The second point, which Suárez conveniently forgets, is that the agricultural workers themselves put forward demands from January 1959 onwards, mainly the demand for 'work or land' a demand which had revolutionary implications. It must be remembered that seasonal and even permanent unemployment was high in Cuba, the average rate between 10 and 15 percent, the seasonal peak for agricultural workers reaching perhaps 50 percent. In so far as the labourers demanded assured work, they were behaving as proletarians. When they demanded that the state should take over the farms or estates in order that they should have work assured they were again behaving as proletarians. When they occupied latifundia they were this time behaving as peasants in a jacquerie. There is evidence for all these types of conduct during the period from the spring of 1959 to the end of 1960.

Draper, discussing Huberman's and Sweezy's *Cuba, Anatomy of a Revolution,* says,

> For Marx, the notion that the peasants would have been the driving force of a socialist revolution would have been simply unthinkable . . . The alleged role of the working class [which Huberman and Sweezy later introduced] in this revolution is just as fanciful as that attributed to the peasantry. (1962:45)

Draper derides Huberman and Sweezy, saying that they

> discovered via a translator that Cuban peasants do not want their own land, they did not even understand the question of owning their own land 'until it had been repeatedly rephrased and explained' . . . 'if so [Draper remarks] the Cuban peasants are truly unique . . . (ibid.:34)

Draper then discovered, apparently by introspection, that peasants wanted only land, and on the other hand he thinks that this wish did not make them into the driving force of the Cuban revolution. I myself think that labourers wanted land *or* work; this was also true of some sharecroppers. Neither land nor work was made automatically available to labourers by the land reform law of May 1959, because it was a very moderate law.

I shall not go into a full account of the contents and ambiguities of the land reform law. Suffice it to say that it guaranteed neither assured work nor land to the half million labourers;

it would have meant, if literally applied, little difference to
most of them since it allowed landowners and large tenants to
keep at least a thousand acres. Except in Camagüey, where
there was much pasture and little agriculture, there were few
landholders, especially in the densely populated areas, whose
farms exceeded this generous limit — which was a minimum,
and could still be exceeded if the Agrarian Reform Institute
thought fit, as it frequently did. In Camagüey, Oriente, Pinar
del Río, Isle of Pines, and some regions of Las Villas, 50 caballerías, and not 30, was the limit for pasture land, i.e. 1,600
acres. One must be careful and not mistake ownership and
landholding under a tenancy title. It is true that American
and Cuban sugar cane companies owned enormous amounts
of land; but this land was in the hands of colonos, large,
medium and small.

The crucial issue, therefore, was what was to become of
this class of colono farmers, especially the medium-sized and
large ones. The law said that those who farmed land belonging
to sugar companies, or to other landlords, would be entitled
to buy up to 30 caballerías. Subsequent legislation which
never appeared would have determined the procedure. It
would seem that the revolutionaries in May 1959 were thinking that a land reform which got rid of the American companies' landed properties and gave ownership of the land to
farmers and small tenants was good enough.

In the expropriated portions over the limit of 30 or 50
caballerías cooperatives were to be formed. The land was not
to be distributed to agricultural workers, Fidel Castro explained, so that the government would not be accused of
creating minifundia. In any case, the amount of land available
would have been small — probably less, had the law been applied literally, than the amount of land farmed by cooperative ejidos in Mexico — hardly a socialist triumph. The government thought rather in terms of making more agricultural
land available to cooperatives through drainage of marshes and
ploughing up pasture land. That Fidel Castro was thinking
along these lines is shown by the amount of time and effort
he spent in the marshland of the Ciénaga de Zapata during
the first months of the revolution — not exactly the behaviour
to be expected of him had he been interested in rousing the
rural rabble of the sugar lands.[5]

There is, nevertheless, the theory which, while admitting
the moderation of the land reform law, maintains that Fidel
Castro intended from the beginning to go beyond its bounds.

Moderation was, then, a deliberate deception. This has been argued by Rufo López-Fresquet, a liberal, who was at the time Finance Minister, and who resigned in March 1960. He says:

> This law, if legally applied, would have affected only the owners of some latifundia . . . but [it] was set aside to make way for the agrarian reform that was finally effected, the unwritten one that Castro always had in mind, the one that dispossessed the landlords and gave nothing to the peasants, the one that destroyed the institution of private property and made the State the sole owner of all property . . . (1966:115)

Perhaps Fidel Castro had been all along a secret collectivizer, and perhaps he was expecting that pressure from the labourers against his moderate land reform law would make the law unworkable. The fact remains that the law was a moderate one, and that the peasantry thought it was so and brought pressure for a more radical reform. It was, in fact, a middle class reform law.

If the land reform law, as it stood, had been strictly applied, not much land would have been available to the state for the settlement of labourers. There is evidence to show that the labourers did not wait to be offered a job in a cooperative created in far-away Camagüey; anyway, only a few could have got jobs there because of the investment in housing and ploughing up of pasture land which would have been required and because there were not so many estates even in Camagüey which exceeded by a large amount the limit fixed for the province. Labourers felt, once the zafra of 1959 was over and unemployment grew, that they had a right to get land *in nearby colono land.* Or, if not land, they felt they had a right to have work every day; this is especially important because it was an apparently reasonable demand which did not require revolutionary convictions.

Thus, the Agrarian Reform Institute began to get letters, many of them coming from local trade union leaders of sugar cane farms, asking the state to take over their management — irrespective of whether such farms exceeded the 30 caballería limit. For instance, from a small settlement in Oriente comes a demand for 'urgent intervention of those sugar cane farms in order to make available sources of work which would alleviate the reigning poverty, because the firm does not have the cane fields weeded'.[6] There are quite a few such demands backed by complaints against colonos who did not weed the sugar cane to the necessary extent, in the unemployed labourers' eyes, and had never done so. Similarly, the landowners had never cleared bush from pasture land to the necessary extent; *chapear los potreros* goes the Cuban expression. The generic distinctive expressions are *dar condición a los campos* in order to *abrir los trabajos*.

The fact that landowners during 1959 and 1960 were having a difficult time in trying to get credit was not taken into consideration by the labourers. The credit mechanism was disrupted by uncertainty, especially the uncertainty as to who was was really a *malversador*, owning property subject to confiscation for misappropriated assets. Banks were careful in giving credit. Landowners and farmers therefore had no money to pay for agricultural operations which could be put off for a while. Moreover, labourers were successful in claiming increases in wages or arrears due, or thought to be due, to them.

The amount of work which can be used up in weeding in a climate such as Cuba's is, of course, very great. But similar complaints from agricultural workers seem to be common in other countries where wage-labour latifundism exists, that is, a system of land tenure and use of labour in which moderately large farms are cultivated by labourers, using mules or oxen or simple manual tools, and where cash crops are grown. In fact, I think there is something here which is rapidly approaching the status of a law: employment is given by landowners and farmers according to considerations of marginal productivity; labour is paid wages (or piece-rates) which are above its 'equilibrium' price as they remain stable and must remain stable in the presence of unemployment or underemployment; therefore there is available labour which is not used despite the fact that it could make a modest contribution to output. In the eyes of the landowners, it just does not pay to employ this labour — it is *incosteable*, as they repeatedly told the Agrarian Reform Institute. Accusations of absenteeism, of lack of entrepreneurial spirit, to landowners are beside the point, although very common. The conflict arises precisely because of landowners' and farmers' entrepreneurial employment policy.

Whatever the Cuban landowners' and farmers motivations, the fact is that there was a reservoir of labour in Cuban agriculture which landowners and farmers did not tap; or could not tap, because it would not pay to do so, because they would have been 'exploited' in the Pigovian sense.

The question then arises, why did landowners and farmers

not let or sublet land to labourers in the past, turning labourers
into tenants? The extra output coming from the extra employment would presumably partly accrue to employers willing to
become *rentiers*. And in Cuba, as in other countries, one finds
yet another regularity: legislation providing very considerable
security of tenure and limiting rents had been enacted in the
thirties, forties and fifties and it had had self-defeating effects.

Thus, the greater part of the Cuban peasantry, increasing in
absolute numbers, remained proletarian. The reluctance of
landowners and farmers to let or sublet land is also explained
by the loss of face involved in granting land to labourers; the
attack on latifundia is carried out usually in terms of the doctrine of the 'social function of property', a principle incorporated into the 1940 Cuban constitution and which provided
the rationale behind the giving of ownership title to all Cuban
tenants in 1959. This doctrine is directed against the rentier
owners who fulfil no social function and therefore it implies
owner operation, whatever the economic penalty and whatever
conflicts may then arise with the labourers over unemployment.
If tenants can never be evicted, if rents and sharecropping
arrangements are regulated by law, if landowners lose face by
becoming landlords, then of course landowners and farmers
will resign themselves to farm with hired labourers.

Some of the tasks which unemployed labourers feel ought
to be performed may well prove *incosteable,* too expensive,
for landowners and farmers, and more so if trade unions are
important, as in Cuba, because wages are then likely to be set
institutionally at points above the level they would otherwise
reach. For a liberal-left government, such as the Cuban one in
1959, the situation is most disconcerting. They could not agree
to the wastage of labour. On the other hand, by themselves,
they had not planned to take over the holdings of landowners
and farmers who were entrepreneurially minded; in fact they
had been led to believe by the experts that poverty and unemployment came from a lack of entrepreneurship and also from
the alliance between the rural and urban bourgeoisie who lived
by importing food and other goods from the United States.
But the government was subject to pressure from the unemployed labourers.

In 1959 and 1960 the labourers saw that a few of them,
very few, got jobs in cooperatives. They also saw that tenants
of various types ceased to pay rents in money or in kind and
were being given the ownership of the land they occupied.
They saw that some labourers who, in the past, had been

evicted from plots of land were now re-installed in those plots.
In 1959 and 1960, therefore, labourers were seeing that practically everybody got land provided they were not true full-
blooded agricultural labourers. Hence, letters to Fidel Castro
stating

> The workers' delegate of the Aljovín farm . . . demands on
> the workers' behalf the 'intervention' of all 49 caballerías
> because they are manoeuvring to give them the 19 caballe-
> rías to which they are entitled in a stony part of the farm,
> the worst part . . .[7]

They could not believe that this was actually the result of the
literal application of a land reform law heralded as the solution
for Cuban agrarian problems. Landowners and farmers had the
right to choose where they wanted to keep their thirty caballerías and of course they chose the central buildings and the
better soils.

It is not surprising therefore if the government, facing this
pressure from the labourers, went somewhat beyond what the
land reform law might have led one to expect. In this they had
the help of a useful institution: *intervención,* which did not
originally mean expropriation or confiscation but merely the
taking over by the state of a firm, very often because of a
labour dispute, and only for a short time. Since the 1930s
there had been over a hundred intervenciones, the first and
most famous being that of the American electricity company
in 1933. There was nothing especially sinister in the demands
to 'intervene' farms or other firms.[8]

Besides the government was determined to end what to
them appeared as the monstrous irrationalities of the Cuban
economy. Many prominent sugar mill owners, such as Julio
Lobo and Suero Falla, had argued for years that restrictions
on the production of sugar had been a mistake. Cuba should
have pushed up production regardless of any short-term effects on prices, to beat all competitors. Everybody believed
that import substitution of edible oils, pulses, maize, and rice
could be increased further than it had already been. The problem of what to do with the mass of the unemployed had been
discussed time and again; there was the school of those who
thought that there was room and capital to give them work in
industry and services — such as Pazos in the 26th July Movement's Economic Theses — and those who already said that
only 'agrarization' could absorb the unemployed.[9]

When labourers started to claim work or land, and for the
state to 'intervene' farms on the grounds that they had weeds

or bushes, the government was probably surprised or, at least,
not necessarily pleased. I do not see any other way for them —
except some measure of repression — than the one they took.
The government, or rather the regional departments of the
Agrarian Reform Institute, began to use the device of intervention against farms not always included in the provisions of the
law. After January 1960, when the Supreme Court made available a very ample interpretation of the law, they could say
that the law was not being altogether disregarded. I doubt whether the government themselves were convinced by this casuis-
try.[10] It would seem that they lost almost all of their initial
interest in ruling by law and in private enterprise. Because
they had enacted a very moderate agrarian reform law, they
had to give up these two liberal props if they wanted to keep
the loyalty of the labourers.

Even then, changes came too slowly for the labourers. In
the papers of the Agrarian Reform Institute there are at least
forty instances of invasions of estates by groups of labourers
acting on their own initiative. This despite the severity of Law
No. 87 of February 1959, which declared that anyone doing
his own land reform and taking land under his own steam
would lose the right to get land or to work in a cooperative.
Suárez's interpretation of this law seems unconvincing; he says
that it could be enacted because there was no pressure on the
land. I would say that it *had* to be enacted because there *was*
some pressure. Suárez gives what he claims to be the single instance of public disorder in this connection in San Luis,
Oriente (1967:34, 51). López-Fresquet tells how Fidel Castro
was very annoyed by it; López-Fresquet concludes that his
anger was genuine because he had already plotted to socialize
all land (1966:162–3). Thus we have about forty instances of
of illegal occupations of estates by groups of labourers;[11] not
enough perhaps to talk about a revolutionary situation or
about a peasant insurrection, but surely enough, in such a
small country to conclude that the peasantry played some role
in pushing the revolution to the left and thus 'betraying' the
expectations of the rural middle class.

Apart from the invasions, there are many successful and
some unsuccessful petitions for interventions coming from
groups of labourers; there are also many letters from individual
labourers to Fidel Castro asking for land — I have seen some
five hundred, which is a remarkable number of letters since
most labourers could not write — and there were hundreds
of extra-legal interventions: the local delegates explained to
the Havana authorities that they were compelled to intervene,

regardless of what the land reform law said, because there were
so many unemployed labourers around. There was no peasant
revolution before 1959; but there was a risk of one later on
because the unemployed labourers felt deceived when the promises implicit, in their eyes, in the propaganda on land reform
as a solution for all evils failed to materialize.

Let us now study some of these demands from labourers. I
have said that labourers unemployed during part of the year
wanted work or land: that is to say, when they demanded land
it was in order to have assured work. Demands coming from
groups of labourers or from trade union leaders on behalf of
groups of labourers usually ask for the intervention of a farm
on the grounds that there is work to be done but that the landowner or farmer is not having it done. This is, so to speak, a
perfectly proletarian position; in fact, they were asking for
what amounted to the socialization of the means of production, although they never phrased it in such a way. However,
I found a demand from a local trade union leader who, in the
most parochial spirit, asked for land in order to grow *frutos
menores* — that is, subsistence crops, root crops, maize and
beans.[12] Many demands come from individual labourers who
ask for land in order to be able to feed their families or in
order to keep a cow or in order to build a little house. Here
geographical factors help to explain this way of thinking. The
habitat is dispersed in rural Cuba and a good diet — a sort of
Irish diet — can be obtained from milk and root crops (cassava,
sweet potatoes, etc.) with very little work indeed. There are a
few demands from labourers which fall into the ultimate dis-
grace of asking for 'land to cultivate because he does not want
to form part of a cooperative',[13] or another 'demanding land
for himself alone'.[14] This way of thinking, supposedly very
typical of peasants, probably derived support at the time from
the government's policy of giving ownership title to all small
tenant farmers. Some labourers had been able to grow subsistence crops in small plots during the off season. The Agrarian
Reform Institute refused to give titles to such 'semi-proletarians'
unless they had cultivated more than one-fourth of a caballeria,
on the grounds that minifundism was an evil. Pressure for land
also came from this sector of the peasantry.[15] A typical complaint from a colono farming seven caballerías stated that
there

> was a labourer who sometimes is paid by results . . . and
> who has lived on the farm since 1952, always as a labourer,
> as I may prove from the social security receipts and pay-

roll sheets which I am obliged to keep . . . to whom I gave
a piece of land to cultivate subsistence crops for himself
and his family, but with whom I have never had any sort of
business relation apart from the relation between employer
and employee . . . he is now trying to convince the authorities . . . that he has a right to the land since he works it. I
will not deny that he works the land. As a worker, he has to
work if he wants to earn his wages. But this does not give
him any rights to the land since he is not a rentier, or sub-
rentier, or colono, or subcolono, or sharecropper, nor even
less a squatter.[16]

Neither was he a Rebel Army member, nor had he ever been
evicted. Therefore, as for the majority of labourers, there was
no land reform for him.

On the other hand, however, many labourers wrote to the
Agrarian Land Reform Institute explaining their situation:
'since the sugar crop ended he has worked only four days; he
demands work on any farm, or, if this is not available, land to
cultivate',[17] or 'demanding land to cultivate or work in a
cooperative'.[18]

Two propositions may therefore be accepted. First, that
the idea of getting land was one of the Cuban agricultural
labourers' ideas. Secondly, that some wanted land for its own
sake, and some wanted it to have assured work. There is not
much point in trying to give percentages of how many people
thought in these ways. What is important to notice is that for
the labourers it was not a strange idea to appropriate land belonging to others, and that many said that assured work in a
state-managed farm was as good for them as having land of
their own. Therefore, when the question is asked, as it often is:
are agricultural workers interested in higher wages and assured
work or are they interested in land? — the answer must be that
they are interested in both or either.[19] They can go quite easily,
it would seem, in one or the other direction. To the extent
that they are interested in land, it would be proper to classify
agricultural workers as peasants. But in fact, the labourers'
views could be fairly summed up, in Cuba in 1959 and 1960
as: I would like land, but on the other hand I would not mind
getting assured work; or I would like assured work, but on the
other hand I would not mind a piece of land.

It must be emphasized that in 1959 and 1960 the slogan of
the Cuban authorities was 'land to the tiller', not 'land or work'.
It is true that land expropriated where no former small tenants
were settled was not given to labourers but to cooperatives —

in order, as we have seen, to avoid the charges of minifundism. But even then, it seems that in many cooperatives labourers were given little plots of land. More important, the number of agricultural labourers who got the opportunity to work permanently in the public sector was small until 1961, and continued to be so until the second land reform law in October 1963. After Playa Girón in the spring of 1961 when a great deal of counter-revolutionaries' land was confiscated, approximately two-thirds of the labourers were still employed, or hoping to be employed, by private landowners. In 1962 unemployment was still worrying the authorities very much.[20]

There is, then, some truth in the view put forward by Sidney Mintz: 'a rural proletariat . . . inevitably becomes culturally and behaviourally distinct from the peasantry. Its members neither have nor eventually want land . . . They prefer . . . wage minimums (etc.)' (Preface to Guerra, 1964). However, against this view there are the demands we have examined from labourers asking for land, or asking for land or work. Even as proletarians merely interested in doing away with unemployment, they were led to be interested in getting land or in a change in the ownership of land. Many Cuban labourers, as is suggested by Mintz, were thinking of getting rid of unemployment first; but in order to do so, they developed an interest in the land and views on who should own and manage it. It seems that they saw their own management of it or state management as valid alternatives.

Let us notice that these agricultural labourers, despite their peasant-like interest in the land, put forward deceptively moderate demands: land *or work*. First they moved against unemployment. Then, perhaps, they began to think of equality, of doing away with the difference between manual and intellectual workers, with class differences, etc. Thoughts of the millenium, sometimes assumed to be the characteristic initial views of such people, came some time later. In the beginning labourers were mainly interested in getting a moderate reform: getting rid of unemployment. This reform, if taken seriously, may lead far in a regime of wage-labour latifundia.

It may lead far — in the direction of collectivization, that is — even in countries where the proportion of labourers is lower than in Cuba, and the proportion of peasants higher. This is because among those peasants there are usually a large number of sharecroppers. Cuban sharecroppers, who in 1959 and 1960 seemed quite pleased when the Agrarian Reform Institute gave them ownership of the land, had petitioned the authorities

some years before presenting themselves more as labourers than
as peasants. Thus, they had asked that the government should
force landowners and farmers to pay the minimum wage to
sharecroppers when the value of their share was less than they
would have made working for the legal minimum wage over
the same period. This is a demand which I have found also to
have been put forward by the sharecroppers in Andalusia and
the Po Valley. Yet another regularity: it would be proper to
classify some sharecroppers as agricultural labourers.

The first Cuban land reform was meant to stop the shame
of desalojos once and for all: 'the land to the tiller'. It is not
surprising therefore that many labourers asked the Agrarian
Reform Institute that they too be granted land. This is clearly
shown in the demands of some desalojados for land. Thus the
son of the famous Niceto Pérez, one of the four or five peasant
martyrs who had died in the past defending his right to occupy
a piece of land, wrote to the Agrarian Reform Institute explaining that he wanted some land because he needed a place
to build a little house and because he was working in a quarry
loading stones and he had work only three or four days a
week.[22] It seems fair to assume that if the slogan at the moment
had been 'work or land', many desalojados and some sharecroppers would have asked for work and not so much for land.
Many labourers asked, and still more would have asked, for
land or work. Those who did thereby showed to the authorities that, first, they had to carry out a more thorough land reform than they had initially proposed, and second, that this
land reform could easily result in a socialist pattern of land
tenure.

It took some time to convince the authorities. The first agrarian reform went beyond the limits set in the law, but still left
most land in private hands. The second agrarian reform, in
1963, still left over 40 percent of agricultural land in private
hands.[23] There is now a drive for complete collectivization.
Perhaps because nowadays in Cuba nobody is going without a
job while many people feel they could do with more food from
a plot of land, land hunger has been increasing in the last few
years and social resistance to collectivization is perhaps higher
now than some years before. In 1959 it seems there was more
hunger for work than hunger for land, or rather, labourers
wanted a great deal of land reform as a guarantee for work.
This paramount desire to have assured work could even have
applied to the case of sharecroppers and the like who were
given land ownership instead. Social resistance to collectiviza-

tion was less than it is sometimes assumed, and perhaps also less than the Cuban authorities assumed.

Quite apart from the effect of the changes in economic policy — first diversification, then emphasis on sugar again — the Cuban land reforms have not been conducive to an increase in agricultural production, at least up to now. It is also likely that security of employment has acted as a disincentive to effective work; thus, one is not surprised to find Ursinio Rojas, an 'old' Communist member of the Central Committee and trade union leader, advocating in 1964 something like share-cropping for sugar cane labourers, as a form of incentive: each brigade would earn in proportion to yields in the fields it tended.[24] That the economic performance of state farms is not very remarkable is not surprising. What is worth noticing is that such thorough changes were effected and are now being continued with the complete collectivization of agriculture (complete collectivization being a condition of achieving an equalitarian distribution of goods), and that the country has survived it and the system still holds together.

Let us now rapidly review the reasons for collectivization. I see three main areas of conflict. One is, of course, over the supply of food — peasants sell in the black market or have little incentive to produce because of the scarcity of consumer goods. The second conflict is over the use of land: thus there is now a policy forcing peasants with land around sugar mills to sow sugar cane, keeping only a relatively small area to grow subsistence crops for themselves, and the intention of this policy is to economize on transport costs. It is also forbidden to sow sugar cane together with beans, or rice, or whatever in different rows. Thus, peasants lose all power of decision over their land because the state decides what they should grow on the land, partly with such considerations as increasing exports, or soil conservation or re-forestation, in mind. In practice, I fear, impressive plans to change land use in an area — as around Havana — are sometimes taken, consciously or unconsciously, not so much for economic or geographical reasons as an excuse to take the land, or the power of decision on crops, from the peasants.

The main conflict, as I see it, is that arising over the employment of labourers by the peasants. In 1966 they employed *permanently* 10 percent of agricultural labourers.[25] There are no figures on how many they employed temporarily. It is no use maintaining that socialism in Cuba will mean equality and

non-utilization of material incentives when peasants are ready
to give land to labourers as sharecroppers or even free instead
of money-wages, when they pay them piece-rates for the sugar
cane and coffee harvests and for other tasks, and when they
pay them in food. It is difficult to change these relations unless everybody becomes a state employee. This is what is being
done now: pressure is being brought on labourers not to work
for peasants who are provided with free voluntary or military
labour by the state. Few of the peasants are *formally* expropriated.[26]

Going back now to the situation in 1959 and 1960, we
have seen that the labourers' aspirations to get land, or to get
work on state land, are explained by the existence of unemployment. Landowners and farmers thought it uneconomic to use
this 'wasted' labour. As time went by, one may assume that
this moderate aspiration — land or work — which, however,
had revolutionary implications, was supplemented by the
aspiration towards equality which was born later or, rather,
which later became acceptable. What people appear to think
depends in the long run on what they are allowed to say. Complaints about lack of employment opportunities, and talk on
the prospects for some kind of land reform had been very common in Cuba — equality had been a more dangerous thought
which not even the Communists had dared to entertain since
1934 or 1935.

The Cuban government could have distributed land to the
labourers or could, as it did within limits, have given them work
in state farms set up on expropriated land. It gave ownership
of land to sharecroppers who were not very different from agricultural labourers and who nowadays are being collectivized,
although they were promised they could keep their land inde-
finitely. Even if the more conservative decision of giving land
individually to labourers had been taken, this decision would
equally have alienated the support of the liberals. Núñez
Jiménez, the head of the Agrarian Reform Institute said, in
the summer of 1959, that the land reform meant the introduction of capitalism in Cuban agriculture and the disappearance
of feudalism.[27] Whatever he was trying to convey by this
quaint use of words, it is clear that socialism was far from his
thoughts. But it was the land of the colonos which was in question. It is all very well to defend anti-absentee, anti-imperialist,
anti-feudal land reform; but when governments wish to put
reform into effect they seem to find sometimes that they had

exaggerated the evils of the situation: the land is actually managed by resident landowners or by large local farmers. And it is they who take the decisions as to employment.

NOTES

1. The book first appeared as articles in *Diario de la Marina* in 1927. It was published in English (Yale University Press) in 1964 with a preface by Sidney Mintz. Modern developments were mentioned but under-emphasized.

2. Associación de Colonos de Cuba, Circulares Julio-Diciembre 1936, Anexo a Informe del Comite Ejecutivo (12 August 1936).

3. Besides the sources cited in the previous essay see J. O'Connor (1964: 62–71, 85) and the studies mentioned therein: also M. Sánchez Roca (1944), which includes the text of the Ley de Coordinación Azucarera. As far sugar mill land was concerned this legislation effectively discouraged the growing of administration cane by favouring colonos over administration lands in the allocation of quotas. Discouragement was needed because the colonos had simultaneously achieved a better deal in the payment for cane at the expense of the mills. But, by granting so much security of tenure and fixing such a low level of rents for both sugar cane and non sugar cane growing occupants, this legislation also slowed down the trend towards letting and subletting land to labourers.

4. From the abstract of *Castro and Communism* in the *International Review of Social History*, 1968, XIII:2.

5. See a description of Fidel Castro's adventures in the Ciénaga de Zapata in A. Núñez Jiménez (1960).

6. From correspondence with the Provincial Delegation of Oriente (8 September 1959).

7. From correspondence with Zona de Desarrollo Agrario Havana–7 (12 February 1960): 'Delegado de los obreros de la finca Aljovín . . . pide a nombre de los obreros intervención total de las 49 caballerías porque de las 19 que les tocan, les quieren dar la parte de piedra y más mala de la finca, estimando que es una maniobra . . .'

8. Information on 'interventions' before 1959 is given in *Cuba Económica y Financiera*, (May 1953, March 1954, February 1957 (p 17)).

9. For this debate on development strategy prior to 1959, see for instance *Cuba Económica y Financiera* (November 1954, May 1955).

10. Decisions of the Supreme Court Nos. 7 and 21. Article 48 of the land reform law said that the objective of the land reform was to achieve economic growth, and said that the Agrarian Reform Institute should take the necessary measures to this end. It was argued that 'intervention' of farms in cases not contemplated in the law was therefore not illegal, as it was said to be done in order to increase agricultural output or prevent it from decreasing.

11. Rather over forty references to illegal occupations, some of them mentioning a generalized situation in a whole province. Thus, a telegram was sent in July 1959 to the provincial delegate in Matanzas: 'Investigate occupations of land *por la libre*. Those guilty of infringement will lose the right to the minimum allocation of land. Broadcast this by radio in that province.' (From correspondence with the Provincial Delegation of Matanzas, 16 July 1959.)

12. From correspondence with Zona de Desarrollo Agrario Camagüey 20 (19 October 1959).

13. Provincial Delegation of Oriente (13 July 1960): 'tierras para cultivar ya que no quiere formar parte de una cooperativa'.

14. Zona de Desarrollo Agraria Las Villas—14 (15 September 1960): 'solicitando tierra para el solo'.

15. Correspondence with Zonas de Desarrollo Agraria Camagüey—18 10 August 1960), Las Villas—16 (8 December 1960).

16. Zona de Desarrollo Agrario Matanzas—9 (16 June 1960).

17. Zona de Desarrollo Agrario Camagüey—18 (12 May 1960): 'desde la terminación de la zafra sólo ha trabajado cuatro días, pide occupación en cualquier finca, o en su defecto tierra para trabajar'.

18. Zona de Desarrollo Agrario Oriente—28 (30 November 1960): 'solicitando tierra para cultivar o trabajo en una cooperativa.'

19. One of the many writers on agrarian affairs who asks this question is Professor Henry A. Landsberger in his curiously entitled essay, 'Funcion que han desempeñado en el desarrollo las rebeliones y los movimentos campesinos: Metodo de analisis' (1968).

20. Figures on the numbers of labourers who had got permanent jobs in cooperatives or state farms in A. Bianchi's study of Cuban agriculture in Dudley Seers, ed., (1964:108, 125). For unemployment in 1962 see the speech of Ernesto Guevara to the dock workers (6 January 1962).

21. Acta, 21 August 1942, Asamblea Nacional de Representantes de la Asociación de Colonos de Cuba, pp 50—1.

22. From correspondence with Zona de Desarrollo Agrario Oriente—25 (18 July 1959).

23. Forty-three percent still in 1966, according to *Cuba Socialista* (August 1966:128).

24. In a paper to *Primer Forum Azucarero Nacional. Sección: Organización del Trabajo* (8 September 1964).

25. *Cuba Socialista* (August 1966:128).

26. The preceding paragraphs reflect the situation as it appeared in 1970. The pace towards collectivization has more recently slowed down somewhat.

27. In a speech to the Rotary Club, Havana (18 June 1959) included in *Hacia la reforma agraria* (1960:64).

VI

Socialist Cuba:
Some Economic and Political Questions

Stages in the Socialization of the Economy, 1959–70

The stages in the socialization of the economy have been roughly as follows:

1 1959 – socialization of 'misappropriated assets'. This meant approximately five percent of all capital assets.

2 1959–60 – the first land reform; between thirty and forty percent of agricultural land became state property.

3 August–October 1960 – all large firms, about six hundred, roughly divided half and half between American and Cuban ownership, were socialized (I do not know how many of them had already been 'intervened').

4 October 1963 – the second land reform pushed the share of State ownership of agricultural land to slightly less than sixty percent.

5 March 1968 – all private non-agricultural business still remaining (some 60,000) were socialized or discontinued.

6 Since about 1967 onwards, private agricultural production has been or is being collectivized, by means of compulsory expropriation, formation of mutual aid brigades, etc. Moreover, sons of peasants have no automatic right to inherit their parents' remaining plots of land.

Housing is not included in this list: the urban reform law of 1960 drastically limited private ownership. This list also gives an impression of neatness which does not entirely fit. The Cuban revolution was not a sudden and tumultuous process triggered off by a rising of the masses. For instance, the habit of tipping which in Catalonia reportedly disappeared on the day following July 19 1936, survived in Cuba until 1968 when a campaign of persuasion directed from above during the 'revolutionary offensive' reportedly did away with it. But, on the other hand, the revolution has not always been under the leaders' close control. The process through which many means of production were socialized was *intervención* which originally meant the temporary take-over by the government of a firm, usually because of a labour dispute. The practice became so widespread that the term changed in meaning; thus, in the countryside people speak of the 'first intervention' and the 'second intervention' to refer to the first and second land reforms. There have been many 'interventions' outside the great waves of socialization listed above; as the years went by, mainly because the owners indulged in counterrevolutionary activities, such as going into exile or assumed or real involvement in conspiracy at the time of Playa Girón (April 1961). Thus, although there was no legislation between 1960 and 1968 providing for further socialization in industry and commerce, the middle industrial and commercial sector had by 1968 been eliminated by means of 'interventions' or had gone out of business. That those who remained were small did not mean however that their existence was compatible with socialism, as the Cuban leaders understood the term in 1968. They were 'parasites', said Fidel Castro, 'who lived considerably better than the rest, just by watching other people work' *(El Militante Comunista,* June 1968:8—9). In 1968 Cuba was supposed to be experiencing the 'simultaneous construction of socialism and communism'. To understand what was meant by this formula one must study the debate of 1963—65 on the question of how to manage the economy, and on the question of material versus moral incentives.

This debate of 1963—65 will eventually perhaps not rate very high in the history of socialist thought. But intellectual life in Cuba has never been all that exciting for such a debate not to have had a profound effect on the views of all participants, who included, as writers of articles or at least as readers, most of the top and middle rank leadership of the revolution. It is important to notice that the debate also impinged upon two fundamental questions — the political character of the Cuban revolution (a nationalist mass movement or a socialist revolution?), and on the basic economic problem which has plagued the Cuban economy in the last ten years (i.e. the scarcity of labour for the sugar harvest).[1] I shall take up the first point at the end of this paper. On the second point, had Fidel Castro been willing (or felt able) to take the extreme leftist

course of a much higher degree of equality in work remuneration, this would have surely deprived him of part of the urban
middle class and lower middle class political support (which
he was to lose anyway because of the poor performance of
the economy), but it might have prevented the exodus of rural
labourers to occupations of dubious economic value and the
relative lack of productive drive of those who stayed behind.
If it is true that the problem of the vanishing professional cane
cutter has been the most important single problem of the Cuban
economy during the last years, a leftist, extremely egalitarian
course would have been appropriate not only ideologically
but also from the economic viewpoint. A lot has been done to
dignify manual work, except taking the simple measure of
raising rural wages to a level nearer that of the salaries of urban
bureaucrats — so, in Cuba, as in the rest of the world, it is still
true, despite the triumphs of the partisans of equality and
moral incentives between 1967 and 1970, that the harder
one's job, the less one earns.

Increasingly since 1971 (a period not covered in this essay)
it seems that there are again attempts at introducing 'material
incentives' and that wage differentials are being kept if not increased. This policy is being implemented under the auspices
of the same people who after 1967 were decidedly in favour
of equality and 'moral incentives' (not only Fidel Castro, but
also Jorge Risquet, the Minister of Labour). (By 1975, it is becoming increasing difficult to think that the retreat to 'material incentives' is a tactical retreat.)

The Debate on the Management of the Economy, 1963—65

The questions discussed had to do with the organization of
the economy — centralization or decentralization — and also
with work remuneration — whether workers were to be paid
according to what they produced, on some variant of a piece-
work system based on 'work norms', or whether they were
going to be paid on a more egalitarian basis. Who was going to
decide what was to be produced? Was it to be only the centre
or were the local managers to have their say? Which methods
should be used in order to ensure a willingness to work? These
questions were discussed in the terminology of socialist economics: was the law of value still valid in the period of transition
from capitalism to socialism? What is the nature of the socialized means of production, are they commodities or not? Those
led by Guevara, who thought that in the socialized sector of

the economy there were no longer commodities, were saying
in fact that they did not think that the price mechanism, and
even less the market, should be allowed to play the role that
many in Eastern Europe regarded in the mid and late 1960s
as not at all incompatible with socialism and indeed needed
for an efficient allocation of resources.

There is no coherent account of how the economy was run
when the debate started. The management system widely in
use was the so-called *sistema presupuestario de financiamiento.*
For each branch of industry there was a *consolidado,* a trust,
which got the money needed for their transactions through
the *presupuesto,* i.e. the State budget. Some means of production were directly supplied to the units of production by the
corresponding ministry; they were not bought, nor supplied
on credit, but rather given away. Petitions from the units of
production were screened at the trust or at the ministry level,
but it is not clear what criteria were used. No money was paid
for transactions between the units of production of the same
trust, but money was paid for transactions between trusts and
also, naturally, with the private sector. Money was also used, of
course, for wage payments. Trusts, and their units, were expected
to fulfil the quantitative goals set out in sectorial plans. No economic criteria of success, in the sense of a profit shown in money,
or even in the sense of being able to repay to the Ministry of
Finance the money advances received, was demanded. In labour
relations, some use was made of piece work but in regressive
proportion. Also, *emulación,* in the sense of competition, received great attention and still does; this is the winning of
prizes, such as flags for instance, on patriotic dates, proclaiming
a worker or rather a group of workers to be the most socialist.

A different system of organization was proposed, and to
some extent the proposals were implemented in the agricultural sector; the system of self-financed enterprises, or *autonomía financiera.* They were to be provided with basic means of
production by the state, such as land, buildings, which they
could not sell, but all other exchanges with other state enterprises would involve real buying and selling. They were expected to keep accounts of all transactions, both inputs and outputs, valued at prices laid down by the state or even negotiated between the enterprises themselves. They would get
money from the state banks, and they would have to convince
the banks of the creditworthiness of any project; thus, compliance with the overall plan for the economy was to be en-
sured through the banking mechanism: they were to be bank-

financed more than self-financed. In labour relations, they
would apply a system of piece work.

It was in fact over remuneration of labour where the partisans of decentralization and of material incentives scored some
victories, up to 1966. Managers of state enterprises in revolutionary Cuba have been given only a salary: no system of bonuses etc., was introduced (prior to 1975). But, on remuneration
of wage-workers, and especially in agriculture, a more and
more complex and inegalitarian system was used from 1962
until 1965. Before 1962, the old system had continued, though
with somewhat higher piece-rates and with so-called 'piracy'
of labour between state farms and between the state and private sectors, since unemployment was decreasing fast. The
initially favourable course the debate took for the decentralizers and partisans of material incentives may be seen clearly
in the norms established for the 1965 sugar cane harvest, with
no across-the-board raise for cane cutting wages but with an
individualistic system of special bonuses on top of progressive
piece-work. It does not seem unlikely — though it has never
been said explicitly, to my knowledge — that it was precisely
the introduction of 'work-norms' as a basis for piece-work payment from 1962 on (instigated by the Ministry of Labour and
the Land Reform Institute, presumably against the mild protest from the Ministry of Industry officials) which sparked off
the debate. Such practice seems to have caused some pained
reactions: 'A young revolutionary wrote to me recently to say
that he was horrified that we are introducing piece rates. But
the revolution cannot pay time rates because this encourages
slowness and laziness which are infectious . . . We shall estab-
lish norms and those who work badly will earn less' (Carlos
Rafael Rodríguez, in *Revolución,* 19 June 1962). The contrast
between this attitude, which prevailed until about 1966, and
more recent views (at least until 1971) could not be greater.
Thus, Risquet, the Minister of Labour said in the autumn of
1967 that piece work in the building sector,

> though it undoubtedly was conducive to higher productivity,
> however, with respect to the workers' consciousness, with
> respect to their ideology, this system mechanized the work-
> ers, it based their productive effort on direct material gains,
> and their productive achievements did not have as a cause
> the revolutionary desire to work harder so that all the people
> would have their (economic) problems solved, but rather the
> egotistical desire to earn more, individually, in order to live
> better, individually. *(El Militante Comunista,* March 1968:8)

(Again to draw a parallel with Catalonia in July 1936, it is striking that one of the first measures adopted was the banning of piece work, together with a reduction in working hours.)

Even in agriculture, the proposed centre of economic decision was not the state farm, but the *agrupación*, of which about seventy were created, one for every seven or eight state farms. In the state farms, only expenses for labour are recorded; other inputs and outputs are recorded at the *agrupación* level according to lists of prices given to them. The agrupación is then the unit for accounting, though, as we shall see, no important decisions are taken at this level (this was the situation around 1968). It was expected that in due course the state farms themselves would become financially independent: the reason for not yet giving *autonomía financiera* to them, it was said in 1964, was the lack of trained managers. Later on, as Guevara's ideas won the day, this prospect was abandoned, and even the agrupaciones lost most of the decision-making capacities they had been given. The country's economy, at least to 1971, was run on very centralized lines; there was not a single central plan, but rather a number of 'special plans' on sectorial and/or regional lines: some ninety plans, *Granma* once said, all directly accountable to the top leadership. No mechanism for the co-ordination of these plans seemed to exist other than sudden scaling-ups or scaling-downs of initial goals as unforseen scarcities or abundances developed (of unhulled rice, requiring industrial hulling capacity, for instance). Such a high degree of centralization, and the reversal of the policy towards the use of material incentives in 1966 or 1967 are best understood in the light of the 1963—65 debate to which I now return.

The critics of the centralized system, of whom Bettelheim became the main spokesman, accepted some of the arguments put forward to defend it. Thus, for instance, they accepted that lack of trained managers would make excessive decentralization unworkable. They also accepted that in the Cuban economy which is to a very great extent dependent on imports, it was only at the centre that some grave scarcities became suddenly known (the result for instance of American pressure on potential European exporters), and priorities became so clear that there was not much point in giving accounting prices to such imports, let alone in allowing enterprises to bid in a market. Also, some branches of industry, such as electricity and oil refineries, lent themselves, as everybody agreed, to central administration. In some industries, such as sugar and tobacco, opponents of centralization were also ready to concede that decentralization, if carried out horizontally — splitting up the agricultural side from the industrial side and from the external trade side — could worsen the situation; vertical integration required some large degree of centralization.

The critics of the centralized system drew limits on the extent to which they wanted to decentralize. Thus, Bettelheim argued against René Dumont's proposals to give a large degree of autonomy to the brigades of agricultural labourers. The brigades' remuneration ought to depend, said Dumont, and also, significantly, some of the 'old Communist' trade union leaders, not only on the amount of work performed but also on economic results. Remuneration should take into account the labourers' efficiency in the use of inputs — not only how many bags of fertilizer they spread but also the yields achieved at the end of the year. Labourers would then take care to improve the quality of their work and they might even decide to carry out small improvements and investments using their extra time or extra effort. Bettelheim thought that the importance of the decisions which could be taken at this level did not justify granting financial autonomy and economic responsibility to brigades of agricultural labourers. For him, economic responsibility should be placed on the hands of the 'real economic subjects'.[3]

Bettelheim then argued that property relations, the juridical fact that the state owned a large sector of the economy did not mean that the state had the ability to run the economy centrally. They had, so to speak, bitten off more than they could chew. There was a difference between legal subjects and economic subjects; the criterion for defining economic subjects was the effective capacity to decide. Thus, in agriculture, the 'real economic subject', in the present degree of development of the productive forces, will be located on average at a lower level than in industry, because there are many variables to be taken into account, because these variables are very unpredictable, and because individual observation is still important.

Guevara's reply, supported by Mandel, came on different levels. He wrote that Bettelheim was applying mechanically the law of necessary correspondence between the degree of development of the productive forces and the relations of production. Bettelheim was saying that because the Cuban economy was still rather underdeveloped, many decisions were best taken at a low level. The centre could not have the necessary information, and the scale of operations of most units of the Cuban economy was not large; even if many units were already state property, the state lacked the ability to take decisions concerning them. But Guevara said that the law of necessary correspondence was subject in Cuba to some perversities. It was this law which applied also to the capitalist system as a whole and explained why capitalism was doomed. Guevara said, ironically, that it was quite clear to him that the Cuban economy and society had not yet reached such an advanced state on the eve of the revolution and nevertheless a socialist revolution had taken place, because of the leaders' awareness and because they had been taught by the socialist experiences of the present century. Would Bettelheim deny this was a socialist revolution? He was in danger, Guevara remarked, of taking the same position as those who said that a socialist revolution was impossible in Russia because of the low degree of development of the productive forces. This is one line of attack, which eludes the real question raised by Bettelheim.

The second line of attack against enterprise autonomy goes over ground which Bettelheim concedes in fact as common ground. Lack of managers, extreme scarcity of some imported goods when the American boycott was still causing great difficulties, the low number of industrial enterprises in Cuba — less than in Moscow alone — the good system of telecommunications — sugar mills, scattered, all over Cuba, all have telephone and telegraph links.

The third line of attack I take to be the decisive line. It is indeed possible, as all writers in the debate were perfectly aware of, to have a centralized system and material incentives at the same time. But Guevara's question was: is it possible to have a decentralized system, a system of enterprise autonomy, and have at the same time egalitarianism in the distribution of goods? Guevara thought it was not. Nobody had proposed in Cuba enterprise autonomy in the sense of allocation of resources through market prices, but only in the sense of giving a larger measure of economic responsibility to enterprises (in their field of activity and in investment), prices being still centrally fixed, though playing a role in decision making at the enterprise level. But if the criterion to assess the success of a unit of production was not whether the quantitative goals set to it in the plan were fulfilled, but whether the enterprise paid for itself, then, in Guevara's view, it became difficult not to use material incentives both for managers and workers.

It might be argued that managers could run enterprises under a decentralized system just by playing the rules of the game, as a game, or under the threats and rewards of admini-

strative downgrading or promotion. But it seemed clear to
Guevara that under a decentralized system, where managers
would be responsible for economic results, the similarities between this system and the recently deceased capitalist system,
would make it impossible not to give monetary incentives to
successful managers. The *signo de pesos,* as Fidel Castro was
to say later, would never be erased from their minds. Still
more convincing is the argument that if managers must mini-
mize costs, they, taking a dim view of human nature would
resort to material incentives in the well-founded hope of
lowering unit labour costs. Some of the proposals for decentralization which Guevara's opponents put forward were
indeed pushing things far in this direction, such as the proposals to turn brigades of agricultural labourers into sharecropping groups: under such a system one could expect the
workers to start quarrelling among themselves, the older or
less able being thrown out of the brigades. After all, one of
the reasons for turning the cooperatives into state farms had
been that the cooperatives had divided the workers into cooperativists and casual labourers. In the final analysis, then,
the question of centralization vs. decentralization boiled down
in the Cuban debate to the question of remuneration of
labour, and to the question of whether it was possible to dis-
associate the problem of incentives to work from the problem
of the distribution of goods. Guevara took the line that people
would work because they would feel it was their social duty
to do so even though the distribution of goods was egalitarian
and guaranteed to everybody. Arguments in favour of market
socialism are often based, in the last resort, on the contrary
anthropological assumption.[4]

Guevara was in favour of centralization and against material
incentives. There may be an economic cost to pay for such
policy, on both counts — inefficient allocation of resources,
including labour, and lack of motivation to work because of
an egalitarian distribution of goods. Guevara never admitted
that his policy would mean a lower rate of growth than either
the decentralizers' policy or, even less, than capitalism which
he thought meant, in the Third World, underdevelopment
through imperialist exploitation. But he said clearly that he
did not see in socialism a system which recommended itself
primarily because of better economic performance: 'I have no
interest in a socialism without communist morality. We fight
against poverty, but also against alienation. One of the fundamental objectives of marxism is the disappearance of the in-

dividual search for gain and lucre as a psychological motivation.' By 'alienation' he meant here reliance on the profit-
motive. The main point at issue was well summarized by
Alberto Mora, one of his opponents: 'Emphasis has been
placed on the fact that the use of material incentives and the
criterion of profit as a norm to measure the efficiency of
enterprises become basic motivations in economic life, and
this hinders the development of a socialist consciousness in
man' (*Nuestra Industria. Revista Económica,* August 1965:24).
This was the discussion of 1963—65. The public debate stopped
early in 1965, and it looked as if the proposals of the decentralizers were going to be put into effect. Guevara denounced
the use of international prices for trade between socialist and
underdeveloped countries in a speech in Algiers in 1965. This
was indeed going far. Guevara then disappeared from public
life, to prepare for future guerrilla campaigns.

Signs that the decentralizers had won the day were the complex system of remuneration of labour in the sugar harvest of
1965, and also Fidel Castro's dissimulation when talking on
these matters. Fidel Castro has scarcely ever written for publication. His public pronouncements are usually made in long
speeches addressed to the public at large, and he appears to
speak off the cuff. Dissimulation is, therefore, perhaps too
strong a word, but it is difficult to find a different one. For
instance, he suggested that motorcycles and vacation trips to
beaches for the top cane cutters should qualify as moral incentives; he also suggested that the question of how to remunerate
labour was a short term problem, not very important since the
development of technology would do away with it; finally,
'social consumption' was increasing so much that the question
of whether an egalitarian system of remuneration was needed
for an egalitarian distribution of goods would not arise. Hence,
perhaps, the promise to do away with rents for housing in
1970, when the real problem was of course not rents but
houses. Fidel Castro has often spoken of the day, not far away,
when rationing would disappear; this could be followed in a
later speech by a reference to the day, also not far away, when
money would disappear. Perhaps the expression 'distribution
of social production' will be substituted for 'rationing' at a
later date, when rationing quotas become more abundant. But,
because much emphasis has long been placed on the happy day
on which rationing would be abolished, it will require some
effort at explanation to show the people that there is a third
alternative to distribution *por la libreta* (i.e. rationing) or *por*

la libre (i.e. in the market), which still does not entail unlimited availability. Poor people like rationing, in Cuba as elsewhere.

Since 1966–67 (and at least until 1971), the Cubans shelved the plans for decentralization and for the introduction of material incentives. It would seem, though I am not sure, that Guevara did not leave in defeat but in triumph. Fidel Castro said approvingly in his tribute to Guevara in October 1967: 'he saw with absolute clarity moral incentives as the fundamental lever in the construction of communist society.' After the 'microfraction' affair of January 1968, to oppose the line in favour of moral incentives amounted, for party members, to fractionalist activity. (By 1975, the reverse was true.)

It was not uncommon to hear Cuban cynics remark in 1968 (and no doubt still today) that material incentives had been abandoned because there was nothing material to stimulate people with, such was the state of the economy and the magnitude of investment. Increasing the fund for investment might have been an additional reason for the egalitarian policy, or perhaps a mere side effect.

No money was used after 1967 for transactions inside the state sector, not only inside the same trust but also between trusts. Also, there was a move towards abolishing piece work, not completed in agriculture, and the peasantry (some 200,000 owners, half of whom got ownership rights in 1959–60) was being collectivized. This last was a theme not discussed in the otherwise very open debate of 1963–65.[5] Fidel Castro in the ANAP congress of May 1967 promised again to keep peasant agriculture for thirty, forty or more years, though his speech in the following ANAP congress of December 1971 clearly shows that collectivization is making progress. Individual plots of land of about three hectares are sometimes left to the peasants, but it is not clear whether sons have a right to inherit them, or to inherit the farms of those not yet integrated into the state sector. Once collectivization is well advanced, a speech might come in which Fidel, at great length and with considerable repetition, will elaborate on the obstacle posed by the peasantry to an egalitarian society; he will also say that they were less efficient producers than collective agriculture in an age of technological wonders – this being a moot point – and will insist that the peasants have been the staunchest allies of the socialist revolution, all this sprinkled with remarks on erosion and reforestation and on the education of peasants' children for a less egotistical way of life. Not for Cuba the strategy of attempting to set the workers against the peasants.

It is however the obvious existence in Cuba of this firm proletarian social base which makes collectivization easy to implement, together with the small proportion of peasant owners
in the active population (in the region of ten percent). This
policy has however an economic cost (in production, though
not perhaps on the surplus) which goes some way towards explaining the continuing economic difficulties.

There was something of a paradox in choosing agriculture in
1963 as the privileged sector (which, through sugar exports,
would function as the capital goods sector) and choosing, three
or four years later, the policy of centralization and moral incentives. Perhaps in Guevara's preference for import-
substituting industrial development there was the awareness
that the cost of centralization and of non-reliance on material
incentives would be greater in agriculture. But, while it is true
that some early experiments in industrial import-substitution
failed and have been discontinued, it would be wrong to say
that Cuba is only developing its agriculture (or, more than
developing, investing a great deal merely in order to stay where
they were: at least so far). Nickel mining has grown; fertilizer
and cement plants, milk processing plants have been built.
Conspicuously lacking are the light industries manufacturing
consumer durables which a capitalist country of Cuba's size
and level of income would have developed, and which Cuba
tried to develop in the first years after 1959.
    While the main economic problem seems to have been scar-
city of labour at the peak of the sugar cane harvest, one can
foresee that when mechanization of cane-cutting finally arrives
(loading has been mechanized for the last few years), there
might be a problem of surplus labour force. The economy
must be able to absorb the growth of the labour force — in
size and also in qualifications, through the education programme — at higher levels of productivity. The birth rate (still
nearly thirty per thousand) is coming down, and there is much
scope for employment in the building industry and in the in-
tensification of cultivation. When the time comes to expand
the service sector, the field will be quite clear of the urban pro-
liferation of underemployed men and women which disappeared in the March 1968 'revolutionary offensive'. However,
there will remain the problem of trying to preserve the high
social valuation of manual work which in these years of labour
shortage had made a timid appearance, as the urban bureaucrats
had been induced to work seasonally in agricultural activi-

ties. The labour shortage, as remarked above, was created by
the early transfer of many cane cutters to other sectors including the Army, the police, and private agriculture; their sons
have often not replaced them, having become scholarship stu-
dents. The productivity of those remaining has decreased. In
1970 average productivity per cane cutting day (for profes-
sionals and amateurs) was fifty arrobas; it was perhaps three
times as large before 1959. The stage has not yet been reached
by far where 'everybody is a worker, everybody is a student,
everybody is a soldier', and the economic need for everybody
to be at least a part-time worker will be less and less obvious,
though still remaining socially essential for a nation where
class divisions of this type are expected to disappear.

Some Examples of the Effects of Lack of Enterprise Autonomy

Up to the 1970s, the economy was run on very centralized
lines and with a decreasing use of material incentives. It is worthwhile, therefore, to be a little more specific on the effects of
the lack of enterprise autonomy, and I shall now give some
examples as regards a) field of activity, and b) investment decisions, mainly with reference to the agricultural sector.

The inability of the Ministry of Foreign Trade to provide
the required inputs for some industries or to give detailed instructions on export specifications, led to the creation in 1964
of some vertically-integrated combinats, such as the egg and
poultry combinat or the tobacco enterprise, with good results
on the agricultural side. In the sugar industry, it was also believed that a purely technological definition of the field of
activity — separating the agricultural from the industrial side —
had many drawbacks. Sugar mills should be able to dictate
the conditions for cane production: varieties, where to plant
the cane (since this affects transport costs), when to start cutting the cane (since this affects milling yields). On the other
hand, and against vertical integration, the sugar mills were not
in a position to take over agricultural activities in general, as
regards decisions on land use, and on the use of other factors
such as fertilizers, water for irrigation, labour. Sugar mills
would lack, for instance, information on labour requirements
for crops other than sugar cane. It is clear that there is no way
out of this dilemma, i.e. whether to integrate the sugar industry vertically, necessarily allowing the mills to diversify into
agriculture other than sugar cane, or whether to leave to the

farms the job of allocating land and other factors to sugar cane
or to other crops, at the cost of disregarding the sugar mill's
need for a sufficient and timely cane supply. The way out is
a system of prices known to the agricultural enterprises and
to the mills. The points to be made in this connection are
1) that the Cubans have been well aware of the problems in a
technological definition of the field of activity of enterprises
2) that they have had no criterion to solve such problems.

As regards investments, the size of the surplus and the share
of investment in the national product are, of course, not determined by an automatic mechanism of interest rates, prices and
consumers' choice in any economic system. But decisions are
then needed on the sectors into which investment will be
channelled. Such decisions were very centralized in Cuba, though
Fidel Castro, in the speeches after the failed ten million harvest referred to the need for small investments. No enterprise
could borrow money for current operations, let alone to buy
means of production in order to expand capacity. Each enterprise was given the inputs presumably needed to fulfil output
goals on the basis of technical coefficients known, or thought
to be known, by experience or, occasionally, by experiments.
For instance, for the ten million ton harvest, sugar mills
pointed out in which sections of the production flow they had
bottlenecks and the programme of expansion was geared to in-
stalling new machinery at these points. This was a special plan,
the *plan perspectivo azucarero*.

As already mentioned, in the first years of the revolution,
attempts were made to diversify the economy to a large extent, substituting all sorts of imports. As this was coupled
with lack of effective financial control, it resulted in the most
extraordinary lack of coordination. The reaction to this was
to concentrate investment in a few sectors, which is both a
consequence and a prerequisite of the lack of enterprise autonomy. The decision to concentrate investment was taken
while the debate on enterprise autonomy was still going on;
to some extent it prejudged its outcome. The main sectors
where investment took place in the second half of the 1960s
were the sugar industry, dams for irrigation, ships for fishing,
three or four large factories (cement, fertilizers), two or three
regional plans for citrus plantations, a couple of regional plans
for rice production, and the increase in the number of cattle
by restricted and selective slaughter. This accounted perhaps
for something like eighty percent of total investment, excluding education and other such 'social' investments. Only an
average of ten thousand houses per year have been built.

Decisions on whether to invest in one sector or another are taken with some 'rationality' in the sense that international prices are used, if they apply. The foreign trade sector is very important in Cuba, over thirty percent of production. For instance, a decision on whether a certain amount of molasses was to be used for cattle feed or to produce domestic cooking alcohol could be taken almost wholly on the basis of international prices. Molasses had a known price; if exported, the meat it would contribute to produce had a price in the world market (even if domestically consumed), and both the alternative fuel to substitute for alcohol and the new stoves that would be needed were to be imported. But although Cuba is very much of an open economy, decisions cannot always be taken on the basis of international prices. Even in this case of using molasses for cattle feed, the cost of the pasture to be used as an alternative source of feed, is not really known. Moreover, and this is crucial, Fidel Castro appears to be unhappy about the use of international prices as a guide to decisions on investment.

Another consideration is that such decisions assume that technical coefficients are known and given. It is not clear what incentive a unit of production will have to improve its efficiency in the use of inputs. Even if there are administrative sanctions for heads of enterprises where technological coefficients deteriorate, this does not guarantee the efficient use of inputs. For instance, sugar mills have a series of technological coefficients they must try to improve upon, the most important being the weight of sugar of certain specifications which is extracted for a given weight of sugar cane. The later the cane harvest starts, the higher the sugar yield, and it is therefore in the interest of the mill manager to have a later start and hope that labour for cane cutting will be forthcoming. He does not have to pay for the cane, and the state farms around the mill do not have to pay for labour at a higher rate in full season. The state farms are not subject to financial responsibility, anyway. One needs a measure of efficiency which goes beyond the technical coefficients. A sugar mill has some autonomy — and the provincial sugar enterprise has a little more - in deciding what to do in order to improve technical performance. But there is no way of knowing whether the decisions are the right ones from a general point of view.

Fidel Castro and the Economists

The strictures by Guevara on the prices obtaining in the trade between socialist and underdeveloped countries have been mentioned. And, in a very radical sense, Fidel Castro himself did not approve of this structure of prices. There was an incident at a meeting in Havana in 1969, a scientific congress with foreign participants convened by the Institute of Animal Science. Its director, a British scientist, had done important research on feeding large amounts of molasses to cattle. To make his point clearer he stated in his paper for the congress that 'a social opinion which requires women to be slim, plus increasing evidence that sugar is related to cardiac deficiencies, combine to weaken the prospect for increasing sugar consumption by the human population; thus, it could very well be that the future of sugar cane rests increasingly on its use for cattle.' In Fidel Castro's view, this opinion was unacceptable, not because it was mistaken on the relative prices of sugar and meat, but because this was

> clearly a European argument. The problem of slenderness can be applied to the women in Great Britain, France, Italy, Belgium, Sweden, and many other countries where the needs for and the consumption of sugar have reached a saturation point. But this argument cannot be applied to Asia, Africa or Latin America. Nor can it be applied to thousands of millions of persons throughout the world who are consuming less than eleven pounds of sugar per capita annually, those suffering from malnutrition, from protein and caloric deficiencies, factors which constitute the cause of average life span of under 30 years in many of those regions. Actually we belong to that underdeveloped world, and we identify with its needs.

This does not mean that Cuba is going to produce sugar to give away and remedy caloric deficiencies throughout the world, or that she is going to convert her sugar into cheap protein to remedy protein deficiencies. This does not mean either that the Cubans are going to substitute systematically a nutritional value standard for a price standard in planning their agricultural sector, though to some extent they follow this sensible policy in order to fulfil the rationing quotas. But it does mean that the Cuban revolutionary leaders do not consider the structure of international prices, which reflects, in part, a very unequally distributed purchasing power, really 'rational' in the sense that it is not a guide to production to meet priority needs: they are quite right. *Todo necio/confunde valor y precio,* wrote Antonio Machado, and one cannot but agree.

The needs of the population are known in a general way,

though as economic development proceeds choice will become
still more difficult. Even now, Fidel's priorities are sometimes
peculiar. Thus, production of root crops, a cheap food in
terms of land and labour could to some extent replace imported wheat and wheat flour, has lagged behind presumably
because of Fidel Castro's prejudice against root crops. At the
beginning of the revolution Fidel Castro once said that if the
Americans blockaded Cuba, the Cubans would eat *malanga,* a
much appreciated root, tastier than manioc, which the people
now long for and which they believe to be very good for children. But, according to Fidel, 'in the case of the domestic economy, of domestic needs, we are not guided by market considerations; our production is determined not by how much the profits would be, but by the needs of the population'(speech on
9 February 1970). Economists might get very annoyed at such
simplemindedness, and at the confusion between a price
mechanism and a market. Economic anthropologists might
perhaps be more sympathetic.

The examples which have been given, while they suggest
some of the costs of the chosen policies, do not, on the one
hand, explain adequately the poor performance of the economy up to 1970, nor, on the other hand, do they suggest that
the economy will not grow (even if the policy of centralization
was maintained). In fact, if the economy had not performed
well, the reason was perhaps not so much the lack of consistent
criteria to allocate resources and the lack of enterprise autonomy, but rather the emigration of technicians, the military
effort (especially the continuous war in the Escambray region
in the four or five years after 1961), the radical changes in
development strategy, and the lack of labour at the peak of
the sugar harvest. There have been difficulties in organization
while digesting the socialized sector. However, once some clear
objectives are set, even though they might not be those which
would yield the largest benefits for the same use of resources,
the economy is then bound to make gains, even if achieve-
ments fall a bit short of the goals. Overambitious plans are of
course harmful; at present, after 1970, an alarming fact is the
excess capacity in the milling sector, developed with a view to
ten million ton harvests, while the agricultural sector is able
to deliver cane for harvests of only five million tons or so.

Centralization, in the sense of lack of enterprise autonomy,
implies costs which have been alluded to. It also means that
monumentally large errors are made, of a kind not reducible
to the system's lack of control over the use of resources, or to

lack of incentive to improve technical coefficients, etc. They arise from the fact that a man, or a few men, cannot know much and may fall prey to resident or visiting 'experts'. Errors of high calibre are described by Sergio Aranda, by Huberman and Sweezy, and not least by René Dumont, himself probably too generous with instant advice on agronomic matters — let alone on socio-economic topics — in his earlier visits. Lack of feed-back delays the date at which it is discovered that a mistake is being made. However, a multitude of small mistakes were also made in the early days of the revolution, when improvised managers full of revolutionary fervour and diversifying zeal, and not subject in practice to financial control, acted *por la libre*. There is an element of reaction to this early situation in the process towards extreme centralization. But there is also another factor explaining why centralization was carried so far, and this is the *parti pris* against centralization and financial responsibility and material incentives. This *parti pris* fitted quite well with the diffidence towards economists, whose role would be much larger in a decentralized system; economists harbour professional doubts on the costs of equality. For instance, on July 26, 1968, Fidel Castro made an enthusiastic speech on equality and on the future abolition of money. Some people actually believed that the use of money was to disappear overnight, as the remaining industrial and commercial private sector had done four months earlier. Fidel Castro told his audience that in a short time there would be no wage differentials in Cuba. Low wages would be raised, as production increased, up to the level of high wages and salaries, which would be frozen. This was not too remote a possibility since differentials are not large for the standards of an underdeveloped or of a socialist country, and real purchasing power is further equalized by the rationing of practically everything and by the enormous prices of the small black market. One could have raised the objection of how the mechanism for allocating abilities which wage differentials provide for, would be substituted. This is not an objection to complete equality since people could perhaps go around with a price-tag in their lapels stating their training costs. Such a suggestion would probably qualify as a 'European argument'. This would not be the way at all of making the *signo de pesos* disappear from people's minds. Such a line of thought might be seen as implying doubts about the suitability, or even feasibility, of the plans for equality. (By 1975, such plans had been abandoned.)

Perhaps, then, the diffidence towards economists had been

an additional reason leading to a concentration of decision-making in the top leadership. Even Juceplan, the central planning agency, did little apart from its attempt at collecting statistics (at least up to 1971). A former deputy director of Juceplan, Albán Lataste, a Chilean economist and partisan of decentralization in the 1963—65 debate, has written an account of his experiences. Lataste never attacks Fidel Castro by name, but he describes Fidel's 'special plans' as *bólidos externos* which arrived in Juceplan to destroy the sums carefully drawn up by the economists.

The Political System: Bajar la Orientacion

Had the egalitarian line continued, this would have had effects in politics, as adumbrated in Fidel Castro's speeches after the eight and a half million ton harvest of 1970.

A writer in the debate of 1963—65 made the point that to resort to material incentives would perpetuate, in the best of cases, the existence of two sorts of people, the rulers and the ruled, the leaders, the 'dedicated' men, and the masses of managers and workers, motivated by economic gain. But, if a moral incentives policy is implemented, this also implies, on the face of it, two sorts of men; those who decide what will be produced, and those whose productive efforts and standards of living are decided from above. Egalitarianism, if carried out to the extreme, would mean, however, that the leaders would have a similar style of life to the led, and this in itself would act as a corrective to decisions taken without reference to the felt needs of the people. If the leaders were living four in a room, they would rearrange priorities as between the building sector and the rest of the economy. A similar point was made by President Dorticós in a speech in April 1972 to the staff of the Petroleum Institute; referring to the difficulties in the distribution of fuel for domestic cooking he said:

> One of the comrades may say: 'the plan for distribution for that region of Matanzas or for that other region of Oriente has been underfulfilled by only ten percent, and this is not too much, it does not even reach twenty percent underfulfilment' — but this is to take a bureaucratic attitude, and one must pay attention to the importance of such things in everyday life. That is to say, if in the home of any of us fuel would be lacking for just one day, this would be truly a problem, because we could not have a hot meal, even if we had enough food, if we had beans and meat available.

Egalitarianism did not reach the extreme of President Dorticós having to go without a hot meal at his home. But if a continuous and evidently sincere emphasis was then placed on the value of equality, this was bound to have effects on political life. It was somewhat paradoxical to say, rightly, that 'nobody should be surprised if any manifestation of privilege should arouse the most profound indignation among the masses', and at the same time to allow the use of the widespread slogan, 'Commander in Chief, wherever it is, however it is, and for whatever it is, give orders', which in Spanish sounds a bit less peculiar but not less distressing. Was it possible to recognize that 'among the masses there is a strong feeling of equality' (Are we to deplore this? No!'), and at the same time to enjoy such a privileged position shared also to some extent by Fidel Castro's close aides? Fidel Castro had said that 'it is no longer a question of helping the people to develop their (socialist) awareness, but of having the people help us to develop ours.' The process, of course, did not begin in 1970 but it was then for the first time openly acknowledged (speech, 23 August 1970). (By 1975, Fidel was again talking *de haut en bas.*)

Equality and political democracy do not always seem to go together, precisely because, in the light of the debate of 1963—65, it was understood in Cuba that centralization is a concomitant of equality. Is it possible to increase the participation of the masses in economic management? Anyone who has spent any time at all in Cuba will agree that it is not only possible but most desirable. It is my impression that (at least in 1967—70) nobody dared take any action — say, to weed a field — unless *bajara la orientación.* This expression originally meant 'to give advice'. But *orientación* later meant order, and nobody would take a decision without orders coming from above, usually through Party channels. (Party members are coopted from among candidates, 'exemplary workers' but not necessarily manual workers, elected by their colleagues in their work place.)

Until the 'new man' made a massive appearance among managers, technicians and economists, the policy of moral incentives and equality of distribution seemed to require a certain amount of regimentation and a nearly personal direction of the economy. The real alternative to running the economy as it had been run, through 'special plans' directly accountable to the top leadership, would be to give more power to the managers, with a relatively higher level of education and who still would come therefore from the middle class, and who even if

elected and supervised by the workers would be unlikely to believe fervently in a really egalitarian socialism. How could the
masses run the economy — through workers' committees, for
instance — in a country with a complex economy (large external trade sector, industrialized agriculture, over sixty percent
of urban population) where very few workers have yet had
the time to learn properly, for instance, the rule of three? It
takes quite a few years of evening school to reach this level.
It could of course be tried; the cost might well be higher than
that of a centralized system. One can also well imagine, and
this is perhaps the decisive factor, what the Russian reaction
would be. Lastly, while it is wrong to assume that the socialization of the Cuban economy was born solely from Fidel Castro's
nationalism, and while there was from the very beginning an
important element of pressure from below both against Cuban
and American businessmen, it must be recalled that pressure
from below did not take the form of occupations of factories
and farms by some kind of workers' councils but rather that
of demands for 'intervention' by the state.

The participation of the masses could mean, however, not
a role in a decentralized management of the economy, but in
setting goals for the economy, for instance, by imposing the
decision to increase root crops output. Union elections might
be one way for the masses to have a greater say, but, in the
light of the debate of 1963—65, one would expect it to be felt
more in establishing priorities than in the day-to-day running
of the economy.

The manager who travels around the island enjoying the
privileges more or less necessary to his function - such as a
jeep, or a ticket for a restaurant meal — is liable, should the
egalitarian line revive, to be told that socialism in Cuba
means equality, and this will undermine his political status.
This does also apply to the top leadership. The inter-
dependence between egalitarianism and democracy is not so
difficult to discover that one might think that the leaders are
not aware of it. This is why one would think there was no reason for the shrill alarm about socialism in Cuba sounded by
Karol. For, despite the realities of increased socialization, some
critics have been unkind. Some, because they did not like
them (René Dumont, for instance, and Hugh Thomas, who
warned against collectivization of the peasantry). Some, because they did not seem to be aware of them. Thus, Karol:

    After Guevara's death . . . Fidel Castro decided to give
    more consistency to a policy which, up to then, was based

on announcing Chinese-type slogans while following a
Russian path of development. He then turned the Ten
Million Harvest into a wager on which the Revolution's
honour depended. He gave up, in practice, his Guevarist
velleities and proclaimed the absolute priority of hard work
and discipline. In the logic of the Russian model, he ought
to have organized a great competition (of the Stakhanovist
kind) based on material incentives. But money, already
nearly worthless, ran the risk of being ineffective, and it
was surely not worthwhile to sacrifice a principle cherished
by Che in exchange for such a weak mobilizing force. On
the other hand, in the existing social framework, moral incentives did not produce the desired effect. There was only
one alternative: militarization (Karol, 1970:538—9).

This view seems to me to conflict with the realities of the
socialization of the non-agricultural private sector in 1968 and
of the collectivization of the peasantry, which required courage
because of its economic cost and because it meant going back
on recent promises, caused, one might say, by an initial non-
socialist approach to the agrarian question. These are signs
that equality is taken seriously. While complete socialization
does not guarantee equality, at least it sets the economic conditions for an egalitarian system of distribution.

However, the struggle for equality, and therefore against
bureaucracy, is not easy. One only needs to read in Karol's
book the description of a journey through the island with
party officials to realize how better food, uncrowded travelling, and comfortable lodgings were taken for granted, even
when egalitarianism was the official doctrine. (By 1975,
Fidel justified material incentives and differences in income
by appealing to the 'law' of distribution according to work.)

My interpretation of the Cuban political system - which might
look, after all, as an elaborate excuse for Fidel Castro's concentration of power — does not only contrast with the journalistic
attacks of Karol and Dumont, but also with the sympathetic
academic analysis of Fagen. The Cuban revolution is seen by
Fagen as a nationalist mass movement led by a small elite of
revolutionaries whose job it has been to instill new political
values in the population. Fagen's point of view is summed up
in his opinion that youth is more important than class in the
Cuban revolution. This interpretation depends considerably
upon selecting for study such organizations and mass under-

takings as the Committees for the Defence of the Revolution
and the Campaign against Illiteracy which make the Cuban
revolution a Third World example of political education and
participation. However, should one look at the Unions, or at
the Communist Party (whose *first* congress was in 1975), or
even at ANAP (the Association of Peasants), then it is clear
that participation by the masses has indeed not been very con-
spicuous. As to the political education of the masses in what
Fagen calls the *new* values preached by the leaders — such as
egalitarianism — one needs to explain, and Fagen does not, to
what extent such values came originally from the working
class, including the rural working class.

NOTES

1.   On scarcity of labour, see Pollitt (1971).

2.   The debate took place in a series of articles in *Cuba socialista* and
Nuestra industria, Revista económica. There is a summary by Sergio de
Santis (1965), which I translated into Spanish, restoring the original
quotations, in F. Fernández-Santos (1967). The topic has also been
dealt with by Bernardo (1971), Silverman (1971), Silverman (1973),
Mesa-Lago (1971).

3.   The same concept is used by Joshua (1967) and Gutelman (1967).

4.   Thus, Ota Šik, 'for the majority of people the prime incentive to
work does not, in the given technical and economic conditions, lie in
work itself, but in acquiring the necessary use values produced by
others' (1967:142).

5.   This point is dealt with by Huberman and Sweezy (1969) who did
not realize, however, that collectivization was taking place.

Bibliography

Affonso, A., 1972, 'Esbozo histórico del movimiento campesino chileno', *Revista latino-americana de ciencias sociales* (Santiago de Chile), III.

Aguilar, L., 1972, *Cuba 1933*, Ithaca, N.Y.: Cornell University Press.

Alberti, G., 1973, 'Peasant Movements in the Yanamarca Valley', Proceedings of the III World Congress of Rural Sociology, Baton Rouge, *Sociologia Ruralis*.

Aldama, Domingo, 1862, Informe del hacendado don Domingo Aldama al General Serrano contra un plan de colonización africana. Cited by Guerra y Sánchez (1970:61).

Aranda, S., 1968, *La revolución agraria en Cuba*, Mexico.

Balogh, T., 1961, 'Agriculture and Economic Development', *Oxford Economic Papers*, February.

Bardham, P. K. and T. N. Srinivasan, 1971, 'Cropsharing Tenancy in Agriculture: a Theoretical and Empirical Analysis', *American Economic Review*, March.

Baraona, R., 1965, 'Una tipología de haciendas en la sierra ecuatoriana' in Delgado (1965)

Bauer, A., 1971, 'Chilean Rural Labour in the Nineteenth Century', *American Historical Review*.

Bazant, Jan, 1972, 'Peones, arrendatarios y medieros en Mexico. La hacienda de Bocas hasta 1867', Rome: Ponencia presentada al III Simposio de Historia Económica de America Latina.

Bernardo, R., 1971, *The Theory of Moral Incentives in Cuba*, Alabama University Press.

Best, L., 1968, 'Outlines of a Model of a Pure Plantation Economy', *Social and Economic Studies*, September.

Bianchi, A., 1964, in Dudley Seers (1964).

Blackburn, R., 1963, 'Prologue to the Cuban Revolution', *New Left Review,* October.

Bonsal, P., 1971, *Cuba, Castro and the United States,* Pittsburgh: University of Pittsburgh Press.

Boserup, E., 1965, *The Conditions of Agricultural Growth,* London: Allen and Unwin.

Calle, Rigoberto, 1968, *Producción de ovinos,* Lima: Facultad de Zootecnia, Universidad Agraria La Molina.

Cavestany, Rafael, 1951, 'El Campo en España'. Cited by Martinez Alier (1971).

Cepero, Bonilla R., 1963, *Obra Histórica,* Havana.

Chayanov, A. V., English translation 1966 (first edition 1925), *The Theory of Peasant Economy,* edited by Basil Kerblay, R. E. F. Smith and Daniel Thorner, Homewood, Ill.: Richard D. Irwin for the American Economic Association.

Cheung, S. N. S., 1969, *The Theory of Share Tenancy,* Chicago: University of Chicago Press.

Chevalier, F., 1952, *La formation des grandes domaines au Méxique. Terre et société au xvie – xviie siècles,* Paris.

Chevalier, F., 1966, 'Témoignages litteraires et disparités de croissance: L'expansion de la grande propriété dans le Haut-Pérou au xxe siècle', *Annales E. S. C.,* XXV:4.

C.I.D.A. Ecuador, 1965, *Tenencia de la tierra y desarollo socio-económico del sector agrícola,* Washington, D.C.: Union Panamericana.

C.I.D.A. Peru, 1966, *Tenencia de la tierra y desarrollo socio-económico del sector agrícola,* Washington, D.C.: Union Panamericana.

Cornblit, O., 1970, 'Society and Mass Rebellion in Eighteenth Century Peru and Bolivia', *St Antony's Papers,* XXII, Oxford.

Conrad, A. H. and J. R. Meyer, 1958, 'The Economics of Slavery in the Antebellum South', *Journal of Political Economy.*

Cuban Economic Research Project, 1965, *A Study on Cuba,* Coral Gables, Fla: University of Miami Press.

Dalton, George, 1972, 'Peasantries in Anthropology and History', *Current Anthropology* June–October.

Delgado, O. (ed.), 1965, *Reformas Agrarias en la América Latina*, Mexico.

Diaz Martinez, A., 1969, *Ayacucho, hambre y esperanza*, Ayacucho.

Dobb, M., 1973, *Theories of Value and Distribution since Adam Smith*, London: Cambridge University Press.

Domar, E., 1970, 'The Causes of Slavery and Serfdom: A Hypothesis', *Journal of Economic History*, March.

Draper, T., 1962, *Castro's Revolution*, London: Thames and Hudson.

Dube, S. C., 1955, *Indian Village*, London: Routledge.

Dumont, R., 1970, *Cuba, est-il socialiste?*, Paris.

Epstein, T. Scarlett, 1967, 'Productive Efficiency and Customary Rewards in Rural South India', in Firth (1967).

Espinoza, G. and C. Malpica, 1970, *El problema de la tierra. Presencia y proyección de los Siete Ensayos*, Lima.

Fagen, R., 1969, *The Transformation of Political Culture in Cuba*, Stanford.

Feder, E., 1971, '*Latifundia* and Agricultural Labour in Latin America' in Shanin (1971).

Federación Nacional de Trabajadores Azucareros y Confederación de Trabajadores de Cuba, 1946, *Por el pueblo y contra sus explotadores, Utilidad práctica del diferencial azucarero*, Havana.

Fernández-Santos, F. and J. Martinez, 1967, *Cuba, una revolución en marcha*, Suplemento a Cuadernos de Ruedo ibérico, Paris: Ruedo ibérico.

Firth, R. (ed.), 1967, *Themes in Economic Anthropology*, London: Tavistock.

Foreign Policy Association, 1935, *Problemas de la Nueva Cuba, Informe de la Comision de Asuntos Cubanos*, New York.

Frank, A. G., 1967, *Capitalism and Underdevelopment in Latin America*, New York: Monthly Review Press.

Genovese, E., 1965, *The Political Economy of Slavery. Studies in the Economy and Society of the Slave South*, New York: Pantheon.

Gibson, Ch., 1964, *The Aztecs under Spanish Rule*, Stanford.

Gramsci, A., 1971, *Selections from the Prison Notebooks*, edited and translated by Q. Hoare and G. Nowell-Smith, London: Lawrence and Wishart.

Guerra y Sánchez, R., 1964, *Sugar and Society in the Caribbean*, New Haven: Yale University Press (1970 fifth edition, *Azúcar población en las Antillas*, Havana: Instituto Libro).

Gutelman, M., 1967, *L'agriculture socialisée à Cuba*, Paris: Maspero.

Hammond, J. L. and B., 1919, *The Village Labourer*, London: Longmans.

Hobsbawm, E. J. and G. Rudé, 1969, *Captain Swing*, London: Lawrence and Wishart.

Hobsbawm, E. J., 1971 (first edition 1958), *Primitive Rebels*, Manchester: Manchester University Press.

Hobsbawm, E. J., 1973, 'Peasants and Politics', *Journal of Peasant Studies*, I: 1 October.

Huberman, L. and P. Sweezy, 1960, *Cuba, Anatomy of a Revolution*, New York.

Huberman, L. and P. Sweezy, 1969, *Socialism in Cuba*, New York.

Hutchinson, B., 1959, in Pan American Union, 1959.

Ibarra, J., 1967, *Ideologia mambisa*, Havana: Instituto del Libro.

Isla, 1936, editorial, 19 December (XVI).

Jayawardena, Chandra, 1963, *Conflict and Solidarity in a Guianese Plantation*, London: L.S.A. Monographs in Social Anthropology.

Joshua, I., 1967, 'Organisation et rapports de production dans une économie de transition (Cuba)', Problèmes de planification (10), Paris: Centre d'Etudes de Planification Socialiste, EPHE (mimeo).

Kapsoli, W., 1971, *Los movimientos campesinos en Cerro de Pasco 1880–1963*, thesis, Lima: U.N.M. de San Marcos.

Karol, K. S., 1970, *Les guerrilleros au pouvoir*, Paris.

Kula, Witold, 1970, *La théorie économique du système feodal*, Paris—The Hague: Mouton.

Laclau, E., 1971, 'Feudalism and Capitalism in Latin America', *New Left Review*, 67.

Landsberger, Henry A., 1968, 'Función que han desempeñado en el
    desarrollo las rebeliones y los movimientos campesinos: Método de
    análisis', *Boletín* (Instituto Internacional de Estudios Laborales).

Lataste, A., 1969, *Cuba, hacia una nueva economía política del
    socialismo*, Santiago.

Le Riverend, J., 1966, *La Républica*, Havana: Editorial Universitaria.

Lliteras, J. A., 1936, 'Tres políticas', *Isla* IV, 4 July (see also articles in
    Isla on 21 November, 1936).

López-Fresquet, R., 1966, *My Fourteen Months with Castro*, Cleveland
    and New York.

Macera, P., 1968, *Mapas coloniales de haciendas cuzqueñas*, Seminario
    de Historia Rural Andina, Lima: U.N.M. de San Marco (mimeo).

Macera, P., 1971, 'Feudalismo colonial americano', *Acta Historica*,
    Szeged. 1969.

Marrero, Levi, 1969, *Geografía de Cuba*, New York (first edition, 1950,
    Havana).

Martinez-Alier, Juan, 1965, article in *Anales de Economia* (Madrid).

Martinez-Alier, Juan, 1968, *La estabilidad del latifundismo*, Paris: Ruedo
    ibérico.

Martinez-Alier, Juan, 1971, *Labourers and Landowners in Southern
    Spain*, London: Allen and Unwin.

Martinez-Alier, Juan, 1973a, *Los huacchilleros del Peru*, Paris: Ruedo
    ibérico.

Martinez-Alier, Juan, 1973b, 'Letters from Shepherds (Peru)', *Journal
    of Peasant Studies*, I: 1.

Martinez-Alier, J. and V., 1972, *Cuba, economía y sociedad*, Paris:
    Ruedo ibérico.

Martinez-Alier, V., 1974, *Marriage, Class and Colour in Nineteenth
    Century Cuba*, London: Cambridge University Press.

Matos Mar, J., W. F. Whyte, J. Cotler, G. Alberti, J. Oscar Alers,
    F. Fuenzalida, and L. K. Williams, 1969, *Dominación y cambios en
    el Perú rural*, Lima.

Meillassoux, Claude, 1972, 'From reproduction to production', *Economy
    and Society*, February.

Meillassoux, Claude,1973,'The Social Organisation of the Peasantry: The Economic Basis of Kinship', *Journal of Peasant Studies*, I: 1 October.

Meillassoux, Claude, 1975, *Femmes, Greniers et Capitaux*, Paris: Maspero.

Mesa-Lago, C., (ed.), *Revolutionary Change in Cuba*, Pittsburgh, 1971.

Monzón, M. A., 1958, 'Nuestra defectuosa estructura agraria', *Cuba economia y financiera*, February.

Moreno Fraginals, M., 1964, *El Ingenio*, Havana.

Mörner, Magnus, 1973, article in *Hispanic American Historical Review*, May.

Mousnier, R., 1973, *Social Hierarchies*, London: Croom Helm.

Neale, Walter, C., 1957, 'Reciprocity and Redistribution in the Indian Village. Sequel to some notable discussions', in Polanyi et al. (eds.) (1957).

Neira, Hugo, 1970, 'Sindicalismo campesino en el Perú', *Aportes* XVIII.

Nelson, Lowry, 1950, *Rural Cuba*, Minneapolis, Minnesota: University of Minneapolis Press.

Núñez Jiménez, A., *Hacia la reforma agraria*, Havana.

North, D. C. and R. P. Thomas, 1971, 'The Rise and Fall of the Manorial System: A Theoretical Model', *Journal of Economic History*, December.

O'Connor, J., 1964, *Political Economy of Pre-Revolutionary Cuba*, Ph.D. thesis, Columbia University

O'Connor, J., 1970, *The Origins of Socialism in Cuba*, Ithaca, N.Y.: Cornell University Press.

Oliveira, Francisco, 1973, 'La Economía brasileña: Critica a la razón dualista', *El Trimestre Económico*, Mexico.

Ortiz, F., 1946 (English edition), *Cuban Counterpoint: Tobacco and Sugar*, New York: Knopf (*Contrapunteo cubano del tobaco y del azucar*, Havana, 1940 and Universidad de las Villas, Santa Clara, 1963).

Pan American Union, 1959, *Plantation Systems of the New World*, Papers and discussion summaries of the seminar held in San Juan, Puerto Rico, Washington D.C.: Social Science Monographs VII.

Pearse, Andrew, 1972, 'Peasants and Revolution: The Case of Bolivia', *Economy and Society* I: 3 and 4.

Peña, Carlos F., 1962, Informe sobre salarios de la Sociedad Agricola y Ganadera Algolán S.A. presentado a la Asociación de Criadores de Lanares del Perú, 6 April (to be found under Asociación de Criadores de Lanares in Centro de Documentación Agraria, Lima, in hacienda Antapongo and in hacienda Conocancha papers).

Pérez de la Riva, J., 1970, 'La contradicción fundamental de la economia cubana del siglo XIX', *Economia y desarrollo*, II, Havana.

Perrie, M., 1972, 'The Russian Peasant Movement of 1905–1907: its social composition and revolutionary significance', *Past and Present*, LVII.

Petri, L., 1955, *Le lotte agrarie nella Valle Padana*, Turin.

Piel, J., 1967, 'Apropos d'un soulèvement rural peruvien au debut du XXe siècle: Tocroyoc (1921)' *Revue d'histoire moderne et contemporaine*, XIV.

Pino, G. del, 1948, 'Nuestra actitud ante los colonos', *Fundamentos*, December.

Polanyi, K., Conrad Arensberg and Harry Pearson, (eds.), 1957, *Trade and Market in the Early Empires*, London: Collier Macmillan.

Pollitt, B., 1971, 'Employment Plans, Performance and Future Prospects in Cuba', Cambridge: Department of Economics, Reprint Series No. 349.

Prado, Caio, 1960, 'Contribuição para a análise da questão agrária no Brasil', *Revista brasiliense*.

Primer Forum Nacional de Reforma Agraria, Julio 1959, 1960, Havana (to be found in the Bodleian library, Oxford).

Procacci, G., 1972, *La lotte di classe in Italia agli inizi del secolo XX*, Rome.

Rappaport, Roy, 1968, *Pigs for the Ancestors*, New Haven: Yale University Press.

Romano, R., 1971, 'Sous-développement économique et sous-développement culturel', *Cahiers Vilfredo Pareto*, LXXIV (Geneva).

Ruiz, R. E., 1968, *Cuba, the Making of a Revolution*, Amherst: University of Massachusetts Press.

Sahlins, M., 1972 and 1974, *Stone Age Economics*, Chicago: Aldine and London: Tavistock.

Salisbury, R. F., 1962, *From Stone to Steel*, Melbourne: Melbourne University Press.

Sánchez Roca, M., 1944, *El derecho de permanencia*, Havana.

de Santis, Sergio, 1965, *Critica Marxista*, no. 5 and 6, September — October.

Seers, Dudley, (ed.), 1964, *Cuba. The Economic and Social Revolution*, Chapel Hill: University of North Carolina Press.

Shanin, Teodor, Ed., 1971, *Peasants and Peasant Societies*, London: Penguin Books.

Shanin, Teodor, 1972, *The Awkward Class*, London: Oxford University Press.

Shanin, Teodor, 1973, 'The Nature and Logic of the Peasant Economy', *Journal of Peasant Studies*, I: 1 and 2, October and January.

Šik, Ota, 1967, 'Socialist Market Relations and Planning', in C. H. Feinstein (ed.), *Socialism, Capitalism, and Economic Growth*, Cambridge.

Silverman, B. (ed.), 1971, *Man and Socialism in Cuba*, New York.

Silverman, B., 1973, 'Some Dilemmas of Cuban Socialism', in Ann Zammit (ed.), *The Chilean Road to Socialism*, I.D.S., University of Sussex.

Sindicato Nacional de Obreros de la Industria Azucarera, 1934, *La zafra actual y las tareas de los obreros azucareros*, Havana.

Stokes, E., 1973, 'The First Century of British Colonial Rule in India: Social Revolution or Social Stagnation', *Past and Present*, LVIII.

Suárez, Andrés, 1967, *Castroism and Communism*, Boston: M.I.T. Press.

Thomas, Hugh, 1967, 'Middle Class Politics and the Cuban Revolution' in C. Veliz, ed., *The Politics of Conformity in Latin America*, London: Oxford University Press.

Thomas, Hugh, 1971, *Cuba or the Pursuit of Freedom*, London: Eyre and Spottiswood.

Tullis, F. Lamond, 1970, *Lord and Peasant in Peru*, Cambridge, Mass.: Harvard University Press.

Vázquez, M., 1961, *Hacienda, peonaje y servidumbre en los Andes peruanos*, Lima: Editorial Estudios Andinos, Monografias Andinas no. 1.

Velho, Otávio Guilherme, 1972, *Frentes de Expansão e Estructura Agrária*, Rio de Janeiro: Zahar.

Wachtel, N., 1971, *La vision des vaincus. Les Indiens du Péru devant la conquête espagnole*, Paris: Gallimard.

Warriner, Doreen, 1969, *Land Reform in Principle and Practice*, London: Oxford University Press.

Wolf, Eric, 1966, *Peasants*, Englewood Cliffs, N.J.: Prentice Hall.

Wolf, Eric, 1969, *Peasant Wars of the Twentieth Century*, New York: Harper and Row.

Wolf, Eric, 1972, 'Comment on Dalton', *Current Anthropology*, June–October.

Wood, D., 1969, 'Las relaciones revolucionarias de clase y los conflictos politicos en Cuba', *Revista Latinoamericana de Sociologia:* I.

Zaldivar, Ramon, (pseud.), 1971, 'Elementos para un enfoque general de la reforma agraria peruana', *Cuadernos agrarios*, I, August (Lima).

Zangheri, R., (ed.), 1960, *Lotte agrarie in Italia. La federazione nazionale dei lavoratori della terra 1901–1926*, Milan: Inst. Feltrinelli.

Index